SUE BRADLEY'S

COTTON COLLECTION

SUE BRADLEY'S

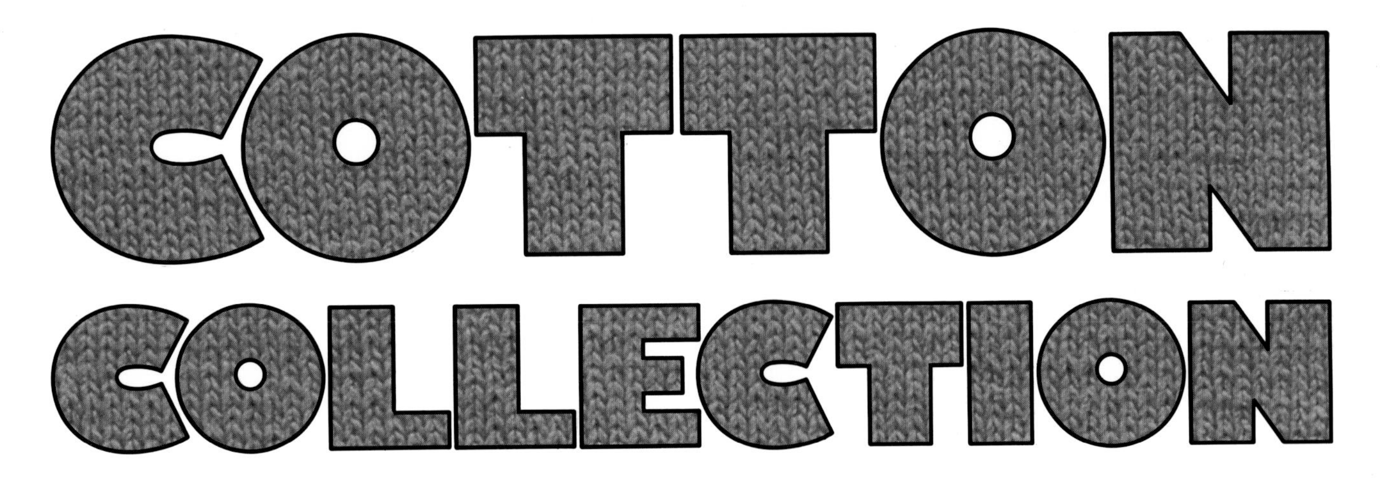

Sue Bradley

HENRY HOLT AND COMPANY
NEW YORK

Published in the United States by Henry Holt and Company, Inc.,
115 West 18th Street, New York, New York 10011.

Published in Canada by Fitzhenry & Whiteside Limited,
195 Allstate Parkway, Markham, Ontario L3R 4T8.

Library of Congress Catalog Card Number: 87-82905
ISBN: 0-8050-0674-5
First American Edition

Produced by Johnson Editions Ltd, 15 Grafton Square,
London SW4 0DQ.

Editor:	Louisa McDonnell
Design:	Anne Fisher
Photography:	Steve Tanner
Charts:	Danny Robins
Pattern checker:	Marilyn Wilson
Proof readers:	Fred and Kathie Gill

With many thanks to the following knitters:
Mary Colbeck
Freddie Craven
Barbara Davis
Rosemary Heath
John Heath
Lorraine Keightley
Betty Lane
Valerie Ruddle
Lila Selman
Phyllis Wilkinson

A special thanks to William Unger for supplying
many of the yarns used in this book.

Typesetting by Fowler Printing Services, London.
Made and printed in Spain by Graficromo S.A., Cordoba.

10 9 8 7 6 5 4 3 2 1

ISBN 0-8050-0674-5

CONTENTS

INTRODUCTION

Cotton is one of the most exciting yarns for the knitwear designer to work with. The uniquely firm nature of the fibre offers endless possibilities for creating well-defined stitch patterns and crisp, bold images which give the garments graphic impact. All sorts of effects can be achieved with the many different weights and textures available; and the enormous choice of characteristically clear colours can be used almost like those on an artist's palette to achieve the exact tone required, whether for the shading on a flower petal, or a bright geometric shape.

In this, my second book, I have designed twenty highly individual sweaters specifically with cotton in mind. A wide variety of yarns has been used, in lots of different colour combinations, and there are designs for every mood and occasion. Lightweight vest and T-shirt styles are perfect for warm summer days, but cotton can also be surprisingly warm, so there are heavier cardigans and sweaters to carry you through to autumn and winter.

The book is divided into four sections, each with its own theme: **Florals**, **Textures**, **Geometrics**, and **Birds and Beasts**. For each garment suggestions are given for two or more different colour variations, shown in knitted swatches, so it is easy to change the colours to your personal favourites.

Although the designs have been made up in cotton yarns, they could be knitted equally successfully in other natural fibres, as long as the yarn chosen is of a similar weight and will produce the same tension. For example a double-knit wool or silk could be substituted for double-knit cotton in any of the patterns to produce an extra-warm or -luxurious sweater.

COTTON CARE NOTES

Cotton has become very popular for hand-knitting in recent years and is still fairly inexpensive in relation to its many unique qualities. It is extremely strong and versatile and produces a naturally soft but durable fabric that is less susceptible to pilling than many other fibres, and is always comfortable to wear.

There are many different types and weights of cotton yarns, from plain hundred per cent cotton in four-ply or double-knit available in hundreds of different colours (and with an optional mercerized finish to give extra sheen) to many more exotic mixes. These include blends of cotton with linen, silk, rayon and acrylic to create interesting slub, twist, crêpe and bobble textures. All these can be used within the same garment, as long as they are of equal weight and can be knitted together to the given tension of the pattern.

If you are not used to knitting with cotton you may find it slightly heavier to work with than wool, since it does not have the same elasticity. It is also more slippery than wool and so when changing colours within pattern motifs it is very important to twist the yarns tightly round each other so that they do not come apart and form a hole. For the same reason it is advisable to sew the ends in very carefully.

When sewing up the seams of a cotton garment you should find it quite easy to obtain a neat, flat seam – oversewing should be unnecessary, as the knitting should be firm enough. A simple edge-to-edge or back-stitched seam is ideal; if the yarn is too thick to sew with easily, unravel it and use a single strand.

Final pressing transforms and perfects the appearance of knitted cotton garments – special attention should be paid to all seams. However, as always, you must follow the instructions on the ball band, since pressing may not be recommended for some cotton-acrylic mixes.

When laundering the garment, you should again consult the ball band instructions. Many cotton mix yarns can be machine-washed and spun-dry, but in general, most hundred per cent cotton yarns should be carefully hand-washed in lukewarm water with a liquid detergent. They should be rinsed very thoroughly and gently squeezed; take care not to lift the wet garment out of the water, as this will cause it to stretch. It should then be carefully reshaped and dried flat away from direct heat (placing between two towels to absorb excess moisture before drying will help speed up the process). If, however, the garment contains a mixture of bright or deep colours, dry cleaning is definitely recommended.

Don't put your cotton sweater away on a hanger – it will stretch and drop. Instead, fold it carefully and store on a shelf or in a drawer. With this careful treatment your sweater will wear well and look new for years to come.

Sue Bradley.

ABBREVIATIONS, NEEDLE CHART AND NOTES FOR AMERICAN KNITTERS

ABBREVIATIONS

alt	alternate
approx	approximately
beg	beginning
cm	centimetre(s)
cont	continue(ing)
dec	decrease(ing)
dk	double knit
foll	follow(s) (ing)
g	gramme(s)
g st	garter stitch
in	inches
inc	increase(ing)
K	knit
K up	pick up and knit
mm	millimetre(s)
oz	ounce
P	purl
P-wise	purl-wise
patt	pattern
psso	pass slipped stitch over
rem	remain(s) (ing)
rep	repeat
RS	right side
sl	slip
st(s)	stitch(es)
st st	stocking stitch
tbl	through back of loop(s)
tog	together
WS	wrong side
yfwd	yarn forward

NEEDLE CONVERSION CHART

UK and Australian metric	Old UK and Australian, Canada, S. Africa	USA
2mm	14	00
2¼mm	13	0
2¾mm	12	1
3mm	11	2
3¼mm	10	3
3¾mm	9	4
4mm	8	5
4½mm	7	6
5mm	6	7
5½mm	5	8
6mm	4	9
6½mm	3	10
7mm	2	10½
7½mm	1	11
8mm	0	12
9mm	00	13
10mm	000	15

NOTES FOR AMERICAN KNITTERS

Both metric and imperial measurements are used throughout the book; American needle sizes are also given, so the patterns should be easily followed by American knitters. However, there are a few differences in knitting terminology and yarn names which are given below.

Knitting terminology

UK	US
cast off	bind off
stocking stitch	stockinette stitch
tension	gauge
work straight	work given

Yarns

UK	US
Aran	fisherman/medium weight
chunky	bulky
double knitting	knitting worsted
4-ply	lightweight

FLORALS

These designs feature flowers, both naturalistic and stylized, translated into clear shapes and fresh, bright colours, all knitted in easy stocking stitch.

The **Bluebell Sweater**, a classic crew-neck style, has a large spray of bluebells growing asymmetrically across a background of crisp white. The **Rose and Check Sweater** combines floral motifs with a chequerboard pattern on an easy T-shirt shape. Small flowers feature in the charming repeat design of the **Fair Isle Cardigan**, while larger-scale blooms decorate the **Button-front Floral Waistcoat**. Pansies in their velvety traditional colours look stunning against the black ground of the **Pansy Sweater**.

BLUEBELL SWEATER

MEASUREMENTS

Three sizes	1st size	2nd size	3rd size
To fit bust	81-86cm/32-34in	86-91cm/34-36in	91-97cm/36-38in
Actual bust measurement	104cm/41in	110cm/43¼in	116cm/45½in
Length	58cm/22¾in	60.5cm/23¾in	63cm/24¾in
Sleeve seam	44cm/17¼in	adjustable for all sizes	

MATERIALS

Yarn

Any **double-knit** weight cotton yarn can be used as long as it knits up to the given tension.
600(650:700)g/21(23:25)oz white (A)
75(75:100)g/3(3:4)oz blue (B)
75(75:100)g/3(3:4)oz green (C)

Needles

1 pair each 3¼mm (US 3) and 4mm (US 5) needles

TENSION

20 sts and 25 rows to 10cm/4in on 4mm (US 5) needles over st st.

> Note
> - When reading charts, work K rows (odd-numbered rows) from right to left and P rows (even-numbered rows) from left to right.
> - When working the motifs, wind off small amounts of the required colours so that each motif can be worked separately, twisting yarns around each other on wrong side at joins to avoid holes.

FRONT

**With 3¼mm (US 3) needles and A, cast on 82(88:94) sts. Cont in K2, P2 rib as foll:
Row 1: (RS facing) K2(3:2), *P2, K2, rep from * to last 0(1:0) sts, K0(1:0).
Row 2: P0(1:0), *P2, K2, rep from * to last 2(3:2) sts, P2(3:2). Rep these 2 rows for 7cm/2¾in, ending with a RS row.
Increase row: Rib and inc 22 sts evenly across row – 104(110:116) sts.
Change to 4mm (US 5) needles and beg with a K row cont in st st working from front chart between appropriate lines for size required.**
Work 110(116:122) rows.

Shape neck

Next row: (RS facing – foll chart) Patt 43(46:49) sts and turn, leaving rem sts on a spare needle.
Complete left side of neck first. Dec 1 st at neck edge on every row until 33(36:39) sts rem. Cont straight until row 128(134:140) of chart has been completed, ending at armhole edge.

Shape shoulder

Cast off 11(12:13) sts at beg of next and foll alt row.
Work 1 row.
Cast off rem 11(12:13) sts.
With RS of work facing return to sts on spare needle, slip centre 18 sts onto a spare needle, rejoin yarn and patt to end.
Dec 1 st at neck edge on every row until 33(36:39) sts rem. Cont without shaping until row 129(135:141) of chart has been completed, ending at armhole edge.

Shape shoulder

Complete as given for first side of neck.

BACK

Work as given for front from ** to **.
Work 14 rows.
Cont in st st and A only until work matches back to shoulder shaping, ending with a P row.

Shape shoulders

Cast off 11(12:13) sts at beg of next 6 rows. Leave rem 38 sts on a spare needle.

BACK AND FRONT CHART

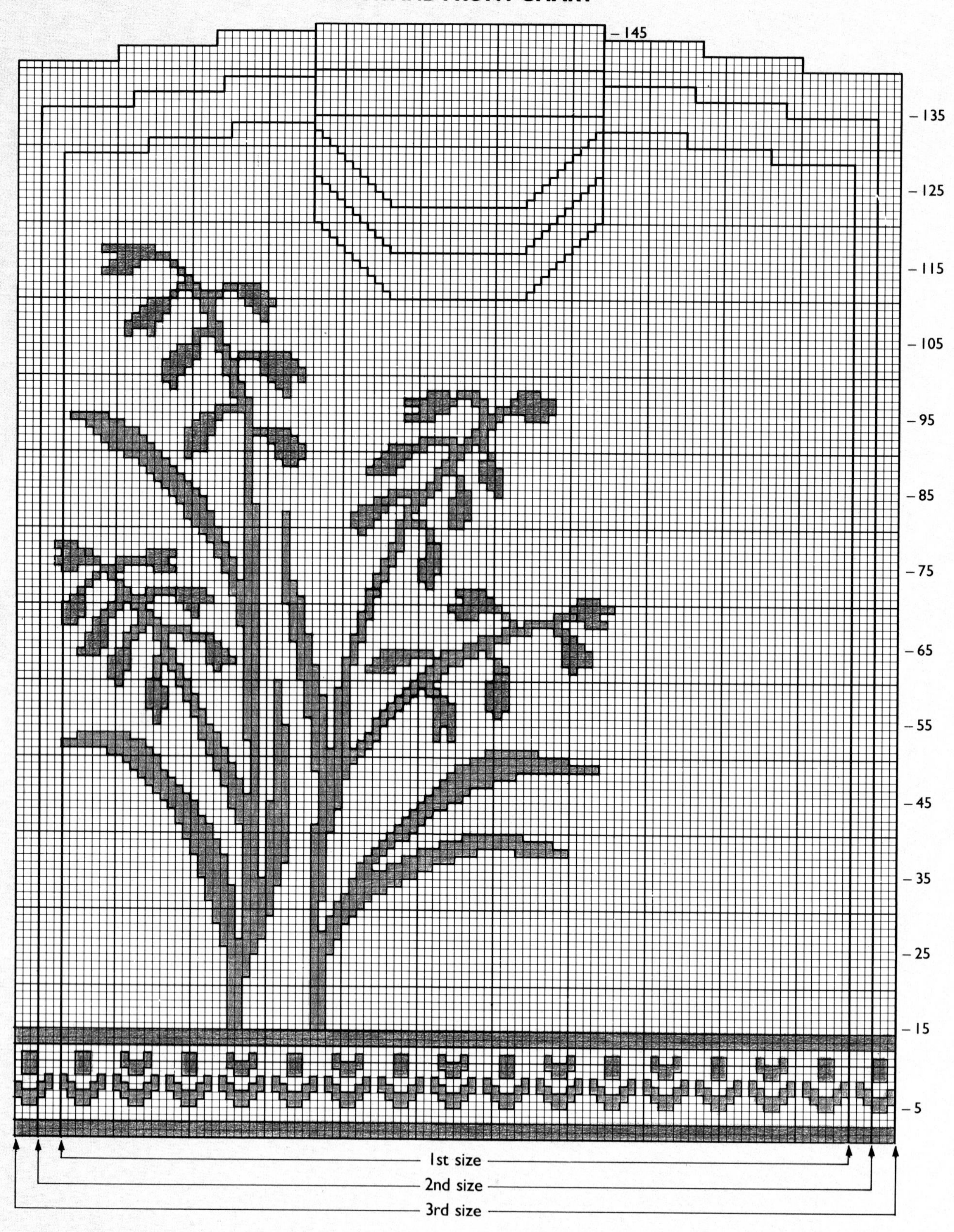

SLEEVES

Make 2. With 3¼mm (US 3) needles and A, cast on 48 sts.

Cont in K2, P2 rib for 6cm/2½in.

Increase row: Rib and inc 24 sts evenly across row – 72 sts.

Change to 4mm (US 5) needles and beg with a K row cont in st st working from sleeve chart, *at the same time*, inc 1 st at each end of 3rd and every foll 4th row until there are 100 sts. Cont without shaping until row 82 of chart has been completed. Work should now measure 39cm/15½in and may be adjusted here if necessary.

Beg with a K row, starting at row 83, cont to work from chart until row 96 has been completed. Cast off loosely.

NECKBAND

Join right shoulder seam.

With RS of work facing, 3¼mm (US 3) needles and A, K up 22 sts down left side of neck, K across 18 sts at centre front, K up 22 sts up right side of neck then K across 38 sts on back neck – 100 sts.

Work in K2, P2 rib for 5cm/2in. Change to B and work a further 2 rows, then cast off loosely in rib.

TO MAKE UP

Press carefully avoiding ribbing using a steam iron or press under a damp cloth.

Join left shoulder and neckband seam. Fold neckband in half onto RS and catch down. Placing centre of cast-off edge of sleeves to shoulder seams, set in sleeves. Join side and sleeve seams. Press seams.

SLEEVE CHART

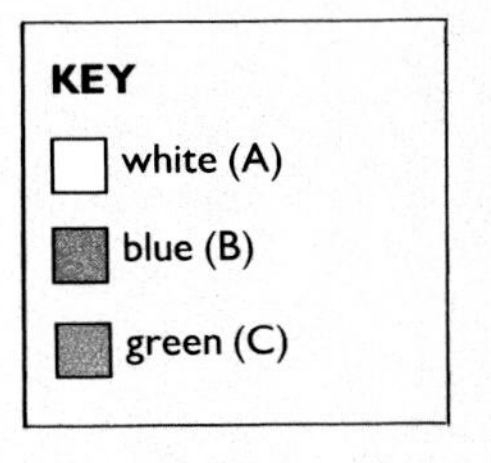

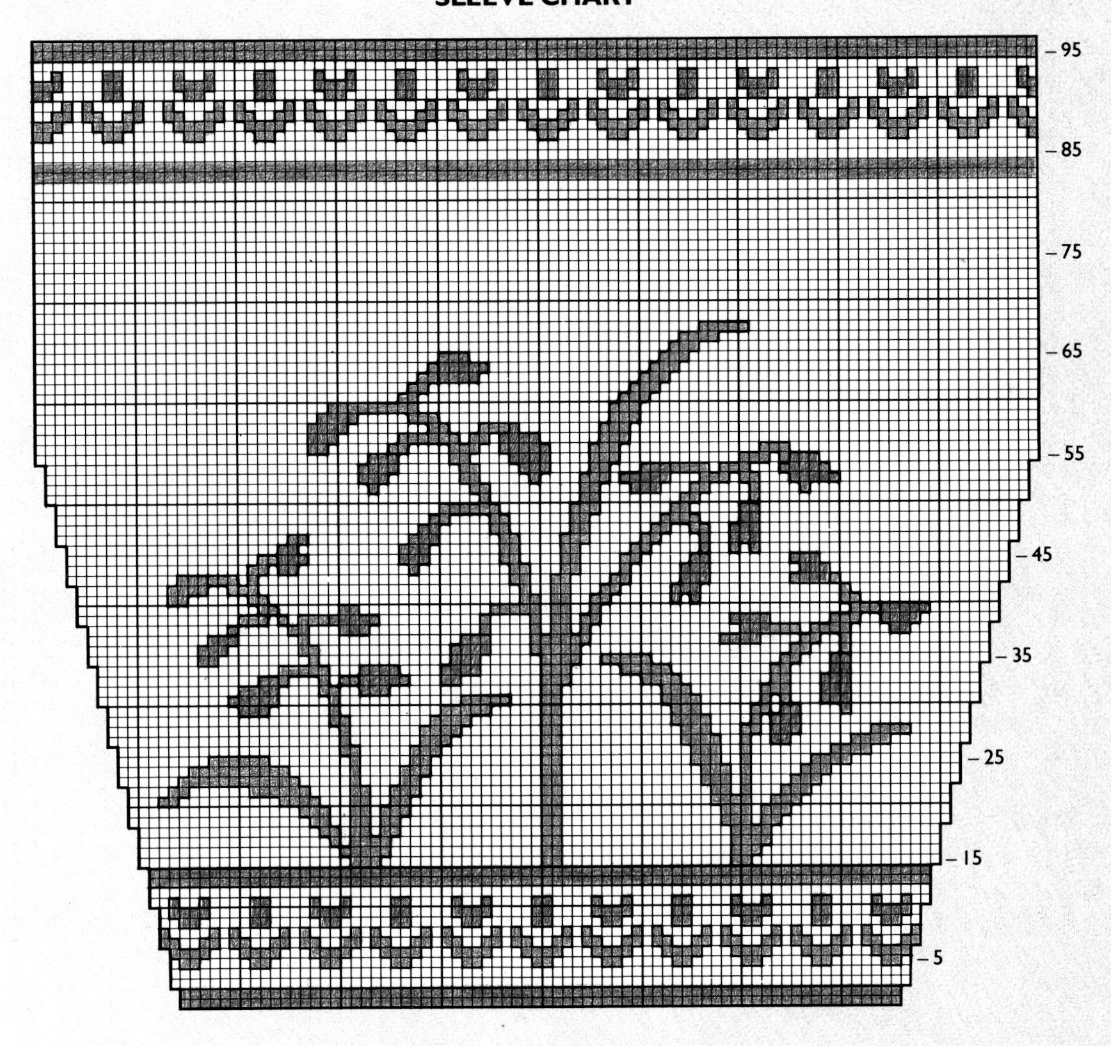

ROSE AND CHECK SWEATER

MEASUREMENTS

Three sizes	1st size	2nd size	3rd size
To fit bust	81-86cm/32-34in	91cm/36in	97-102cm/38-40in
Actual bust measurement	91cm/36in	99cm/39in	106.5cm/41¾in
Length	60.5cm/22¾in	62cm/24½in	63.5cm/25in
Sleeve seam	37cm/14½in	37cm/14½in	37cm/14½in

MATERIALS

Yarn

Any **double-knit** weight cotton yarn can be used as long as it knits up to the given tension.
350(350:375)g/13(13:14)oz white (A)
150 (150:175)g/6(6:7)oz black (B)
100(100:125)g/4(4:5)oz green (C)
100(100:125)g/4(4:5)oz pink (D)

Needles and other materials

1 pair each 3¼mm (US 3) and 4mm (US 5) needles

TENSION

21 sts and 25 rows to 10cm/4in on 4mm (US 5) needles over patt.

Note

- When reading charts, work K rows (odd-numbered rows) from right to left and P rows (even-numbered rows) from left to right.
- When working the rose motifs, wind off small amounts of the required colours so that each colour area can be worked separately, twisting yarns around each other on wrong side at joins to avoid holes.
- On the 2-colour horizontal bands and over zigzag pattern, yarn can be carried on wrong side of work over not more than 3 stitches at a time to keep fabric elastic.

BACK

**With 3¼mm (US 3) needles and A, cast on 76(84:92) sts. Work in K1, P1 rib for 10cm/4in.
Increase row: Rib and inc 20 sts evenly across row – 96(104:112) sts.
Next row: With A, P.
Change to 4mm (US 5) needles. Beg with a K row cont working in st st. Working between appropriate lines for size required work 8 rows of chart 1, 16 rows of chart 2, 8 rows of chart 1 then 18 rows of chart 3. These 50 rows form the patt.**
Rep these 50 rows once more, then work 26(30:34) rows, ending with 2nd row chart 1(6th row chart 1:2nd row chart 3).

Shape shoulders

Keeping chart patts correct, cast off 12 sts at beg of next 2 rows and 10(12:14) sts at beg of foll 4 rows. Leave rem 32 sts on a spare needle.

FRONT

Work as given for back from ** to **.
Rep these 50 rows once more, then work 8(12:16) rows.

Shape neck

Next row: (RS facing – 1st(5th:9th) row chart 2) Patt 36(40:44) sts and turn, leave rem sts on a spare needle.
Complete left side of neck first.
Dec 1 st at neck edge on every row until 32(36:40) sts rem.
Cont without shaping until work matches back to shoulder shaping, ending at armhole edge.

Shape shoulder

Cast off 12 sts at beg of next row and 10(12:14) sts at beg of foll alt row. Work 1 row. Cast off rem 10(12:14) sts.
With RS of work facing, return to sts on spare needle, rejoin yarn, cast off centre 24 sts, patt to end.
Complete to match first side of neck.

SLEEVES

With 3¼mm (US 3) needles and A, cast on 60 sts.
Work in K1, P1 rib for 5cm/2in.
Increase row: Rib and inc 12 sts evenly across row – 72 sts.

CHART 1

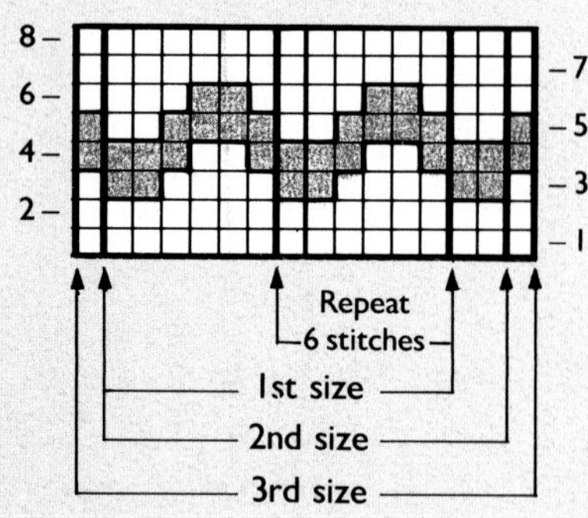

CHART 2

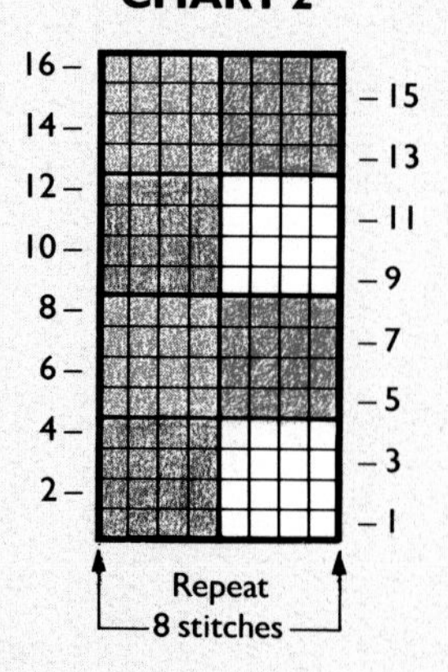

KEY

white (A)
black (B)
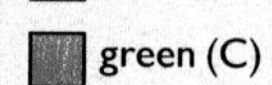
green (C)
pink (D)

CHART 3

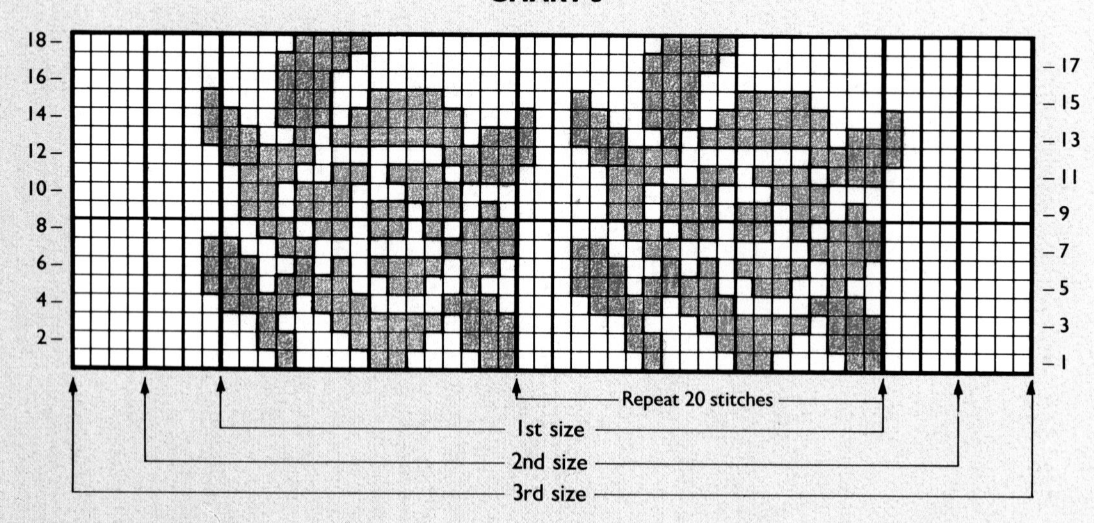

Change to 4mm (US 5) needles. Beg with a K row cont in st st working from chart, *at the same time*, inc 1 st at each end of every 3rd row until there are 90 sts. Cont without shaping until row 56 of chart has been completed, then work rows 33-56 again.
Cast off loosely.

NECKBAND

Join right shoulder seam.
With RS of work facing, 3¼mm (US 3) needles and A, K across 32 sts on back neck, K up 23 sts down left front neck, 24 sts across centre front and 23 sts up right side of neck – 102 sts.
Work in K1, P1 rib for 6 rows. Cast off loosely in rib.

TO MAKE UP

Press pieces carefully using a steam iron or press under a damp cloth.
Join left shoulder and neckband seam. Placing centre of cast-off edge of sleeves to shoulder seams, set in sleeves. Join side and sleeve seams.

SLEEVE CHART

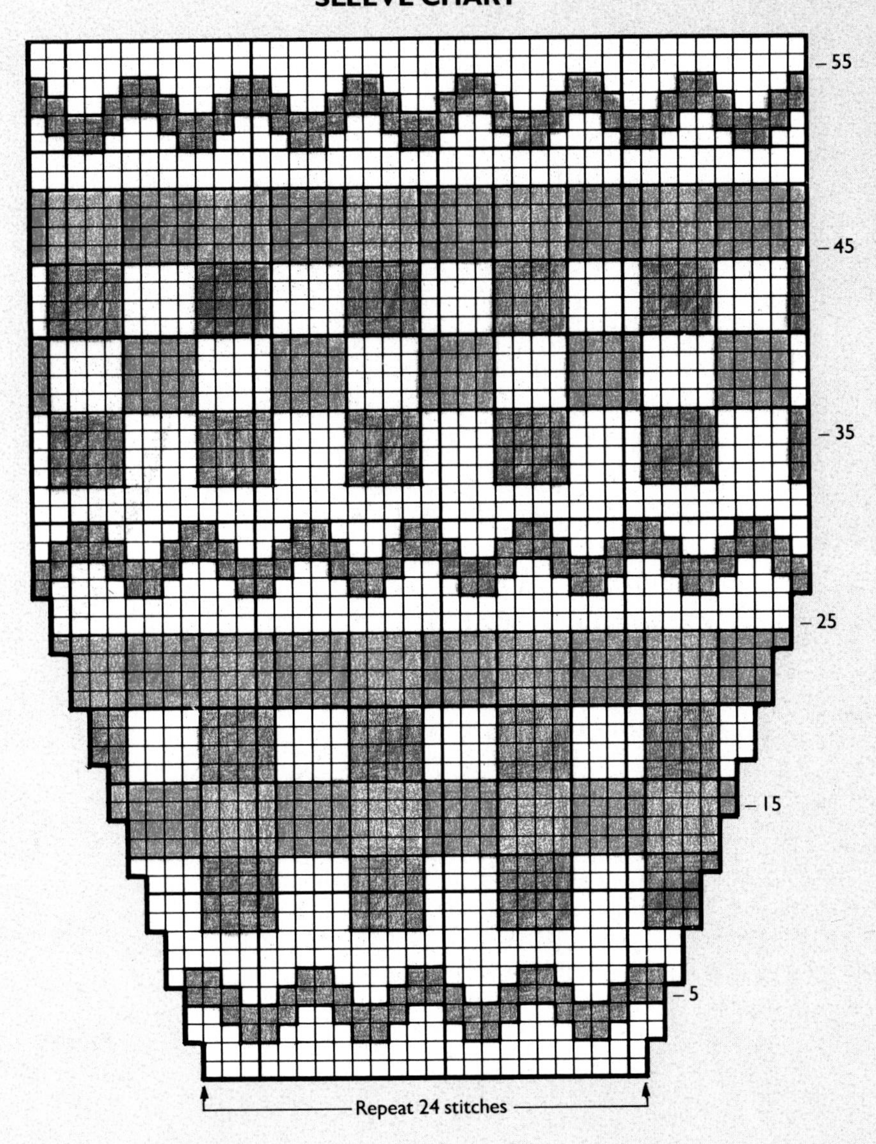

FAIR ISLE CARDIGAN

MEASUREMENTS

Three sizes	1st size	2nd size	3rd size
To fit bust	81-86cm/32-34in	86-91cm/34-36in	91-97cm/36-38in
Actual bust measurement	95cm/37½in	101cm/39¾in	107.5cm/42¼in
Length	66cm/26in	68.5cm/28in	71cm/28in
Sleeve seam	45.5cm/18in	45.5cm/18in	45.5cm/18in

MATERIALS

Yarn

Any **double-knit** weight cotton yarn can be used as long as it knits up to the given tension.
650(650:675)g/23(23:24)oz white (A)
75(100:100)g/3(4:4)oz blue (B)
125(150:150)g/5(6:6)oz yellow (C)
150(150:175)g/6(6:7)oz pink (D)
150(150:175)g/6(6:7)oz green (E)

Needles and other materials

1 pair each 3¼mm (US 3) and 4mm (US 5) needles
3¼mm (US 3) circular needle
7 buttons

TENSION

26 sts and 26 rows to 10cm/4in on 4mm (US 5) needles over patt.

Note
When reading chart, work K rows (odd-numbered rows) from right to left and P rows (even-numbered rows) from left to right.
Yarn not in use should be carried on wrong side of work over not more than 3 sts at a time to keep fabric elastic.

BACK

With 3¼mm (US 3) needles and A, cast on 120(128:136) sts. Cont in K1, P1 rib for 2.5cm/1in.
Change to 4mm (US 5) needles and beg with a K row cont in st st working from chart. Cont until work measures 66(68·5:71)cm/26(27:28)in from beg, ending with a P row.

Shape shoulders

Cast off 13(14:15) sts at beg of next 6 rows. Cast off rem 42(44:46) sts.

LEFT FRONT

With 3¼mm (US 3) needles and A, cast on 60(64:68) sts.
Cont in K1, P1 rib for 2·5cm/1in.
Change to 4mm (US 5) needles and beg with a K row cont in st st working from chart between appropriate lines for size required.
Cont until work measures 56(58·5:61)cm/22(23:24)in from beg, ending with a K row.

Shape neck

Keeping patt correct, cast off 12 sts at beg of next row.
Dec 1 st at neck edge on every row until 39(42:45) sts rem.
Cont without shaping until work matches back to shoulder shaping, ending with a P row.
Cast off 13(14:15) sts at beg of next and foll alt row.
Work 1 row.
Cast off rem 13(14:15) sts.

RIGHT FRONT

Work as given for left front reversing all shapings.

SLEEVES

Make 2. With 3¼mm (US 3) needles and A, cast on 60 sts. Cont in K1, P1 rib for 5cm/2in.
Increase row: Rib and inc into every st across row – 120 sts.
Change to 4mm (US 5) needles. Beg with a K row cont in st st working from chart until work measures approx 45.5cm/18in from beg ending with row 2 of chart. Cast off loosely in A.

POCKET LININGS

Mark the position of pockets on side edges of back and fronts 10cm/4in and 25.5cm/10in above cast-on edge.
Left front
With 3¼mm (US 3) needles, A and RS of work facing K up 36 sts between pocket markers.
Change to 4mm (US 5) needles. Beg with a P row cont in st st, dec 1 st at beg of next and every foll alt row until 22 sts rem. Cast off.

CHART

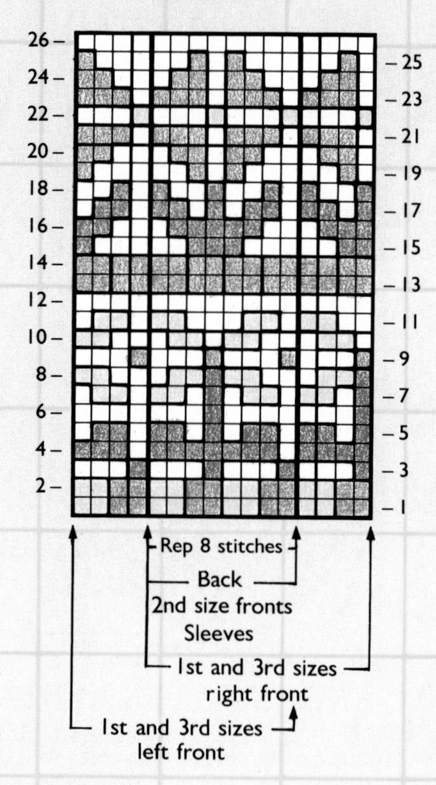

KEY
- white (A)
- blue (B)
- yellow (C)
- pink (D)
- green (E)

Right front
Work as given for left front reversing shaping.
Left back
Work as given for right front.
Right back
Work as given for left front.

NECKBAND

Join shoulder seams.
With RS of work facing, 3¼mm (US 3) needles and A, K up 34 sts up right front neck, 42(44:46) sts across back neck and 34 sts down left front neck – 110(112:114) sts.
Cont in K1, P1 rib for 5cm/2in. Cast off loosely in rib.

BUTTONHOLE BAND

Fold neckband in half onto RS and slip stitch in place.
With RS of work facing, 3¼mm (US 3) circular needle and A, K up 139(145:151) sts evenly up right front edge and neckband.
Work in rows.
Row 1: (WS facing) P1, *K1, P1, rep from * to end.
Row 2: K1, *P1, K1, rep from * to end.
These 2 rows form the rib.
Work 1 more row.
1st buttonhole row: Rib 6, * cast off 3, rib 18(19:20) including st used to cast off, rep from * 5 times more,

cast off 3, rib to end.
2nd buttonhole row: Rib to end, casting on 3 sts over those cast off in previous row.
Work 3 rows rib.
Cast off in rib.

BUTTONBAND

With RS of work facing, 3¼mm (US 3) circular needle and A, K up 139(145:151) sts evenly down neckband and left front edge. Work in rows.
Work 8 rows K1, P1 rib as given for buttonhole band.
Cast off in rib.

TO MAKE UP

Carefully press pieces avoiding ribbing using a steam iron or press under a damp cloth.
Placing centre of cast-off edge of sleeves to shoulder seams, set in sleeves. Join side and sleeve seams and edges of pocket linings.
Sew on buttons.

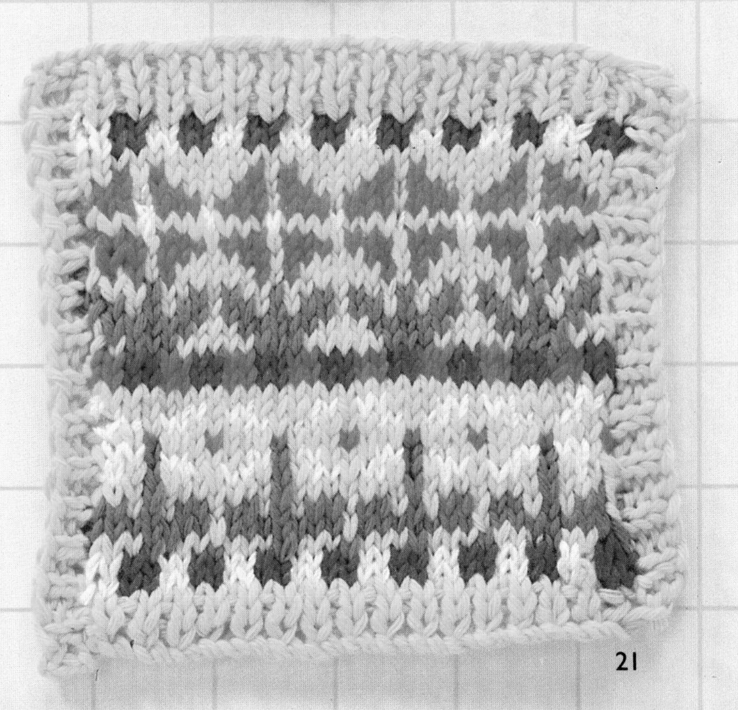

BUTTON-FRONT FLORAL WAISTCOAT

MEASUREMENTS

Three sizes	1st size	2nd size	3rd size
To fit bust Actual bust measurement Length	81-86cm/32-34in 103cm/40½in 52cm/20½in	86-91cm/34-36in 110cm/43¼in 54cm/21¼in	91-97cm/36-38in 116.5cm/45¾in 55.5cm/21¾in

MATERIALS

Yarn

Any **chunky-weight** cotton yarn can be used as long as it knits up to the given tension. One strand 4-ply cotton and one strand double-knit cotton may be used together if preferred.
450(475:500)g/16(17:18)oz blue (A)
25(25:25)g/1(1:1)oz each black (B), pink (C), yellow (D), green (E)

Needles and other materials

1 pair each 4½mm (US 6) and 5½mm (US 8) needles
5 buttons

TENSION

17½ sts and 23 rows to 10cm/4in on 5½mm (US 8) needles over st st.

Note

- When reading charts, work K rows (odd-numbered rows) from right to left and P rows (even-numbered rows) from left to right.
- When working the floral motifs, wind off small amounts of the required colours so that each colour area can be worked separately, twisting yarns around each other on wrong side at joins to avoid holes.

FRONTS CHART

LEFT FRONT

**With 4½mm (US 6) needles and A, cast on 44(46:50) sts. Work in K1, P1 rib for 2·5cm/1in. Inc 1 st at end of last row for *2nd size only* – 44(47:50) sts.
Change to 5½mm (US 8) needles. Beg with a K row, cont in st st working from left front chart between appropriate lines for size required.
Work 64(68:72) rows, ending with a P row.

Shape armhole

Next row: (RS facing – foll chart) Cast off 8 sts, work to end.
Dec 1 st at armhole edge on every foll alt row until 32(35:38) sts rem.
Cont without shaping until row 79(83:87) of chart has been worked, ending with a K row.

Shape neck

Cast off 9 sts at beg of next row, then dec 1 st at neck edge on every row until 10(13:16) sts rem.**
Cont without shaping until row 114(118:122) of chart has been completed, ending with a P row.

Shape shoulder

Cast off rem sts.

RIGHT FRONT

Work as given for left front from ** to ** working from appropriate chart and reversing all shapings so working 1 more row before armhole and neck shaping.
Complete as given for left front.

BACK

With 4½mm (US 6) needles and A, cast on 88(94:100) sts. Work in K1, P1 rib for 2.5cm/1in.
Change to 5½mm (US 8) needles. Beg with a K row cont in st st until work measures same as left front to armhole shaping, ending with a P row.

Shape armholes

Cast off 6(7:8) sts at beg of next 2 rows. Dec 1 st at beg of foll 8 rows – 68(72:76) sts.
Cont without shaping until work matches left front to shoulder shaping, ending with a P row.

Shape shoulders

Cast off rem sts.

NECKBAND

Join shoulder seams.
With RS of work facing, 4½mm (US 6) needles and A, K up 34 sts up right side of neck, 40 sts across back neck and 34 sts down left side of neck – 108 sts.
Work in K1, P1 rib for 2.5cm/1in.
Cast off in rib.

BUTTONHOLE BAND

With RS of work facing, 4½mm (US 6) needles and A, K up 72(74:78) sts evenly up right front edge and end of neckband.
Work 1 row K1, P1 rib.
1st buttonhole row: (RS facing) Rib 3(4:4), *cast off 2, rib 14(14:15) including st used to cast off, rep from * 3 times more, cast off 2, rib to end.
2nd buttonhole row: Rib to end, casting on 2 sts over those cast off in previous row.
Work 2 rows rib. Cast off in rib.

BUTTON BAND

With RS of work facing, 4½mm (US 6) needles and A, K up 72(74:78) sts evenly up left front edge and end of neckband.
Work 5 rows K1, P1 rib.
Cast off in rib.

ARMBANDS

Alike. With right side of work facing, 4½mm (US 6) needles and A, K up 80(82:84) sts evenly around armhole edge.
Work in K1, P1 rib for 2.5cm/1in.
Cast off in rib.

TO MAKE UP

Press lightly on wrong side of work avoiding ribbing using a steam iron or press under a damp cloth.
Join side seams. Sew on buttons.

PANSY SWEATER

MEASUREMENTS

Two sizes	1st size	2nd size
To fit bust	86-91cm/34-36in	91-97cm/36-38in
Actual bust measurement	104cm/41in	110cm/43¼in
Length	57cm/22½in	60.5cm/22¾in
Sleeve seam	35cm/13¾in	35cm/13¾in

MATERIALS

Yarn

Any **double-knit** weight cotton yarn can be used as long as it knits up to the given tension.
675(700)g/24(25)oz black (A),
25(25)g/1(1)oz each lilac (B), purple (C), yellow (D), deep yellow (E), bright green (F), deep green (G)

Needles

1 pair each 3¼mm (US 3) and 4mm (US 5) needles
3¼mm (US 3) and 4mm (US 5) circular needles

TENSION

20 sts and 26 rows to 10cm/4in on 4mm (US 5) needles over patt.

Note

- When reading charts, work K rows (odd-numbered rows) from right to left and P rows (even-numbered rows) from left to right.
- When working the pansy motifs, wind off small amounts of the required colours so that each colour area can be worked separately, twisting yarns around each other on wrong side at joins to avoid holes.

BACK

**With 3¼mm (US 3) needles and A, cast on 84(90) sts.
Work in K1, P1 rib for 9cm/3½in.
Increase row: Rib and inc 20 sts evenly across row – 104(110) sts.
Change to 4mm (US 5) needles. Beg with a K row cont in st st working from back chart between appropriate lines for size required.
Work 52(54) rows.**

Shape raglan armholes

Foll chart, dec 1 st at each end of next and every foll alt row until 30 sts rem, ending with a P row. Cast off.

FRONT

Work as given for back from ** to **.

Shape raglan armholes

Foll chart, dec 1 st at each end of next and every foll alt row until 46(50) sts rem, ending with a P row.

Shape neck

Next row: (RS facing – foll chart) Work 2 tog, patt 14(16), turn leaving rem sts on a spare needle.
Complete left side of neck first.
Still dec at armhole edge as before on every foll alt row, dec 1 st at neck edge on every row until 3(5) sts rem.
Keeping neck edge straight, cont to dec at armhole edge as before until 1 st rem. Fasten off.

BACK AND FRONT CHART

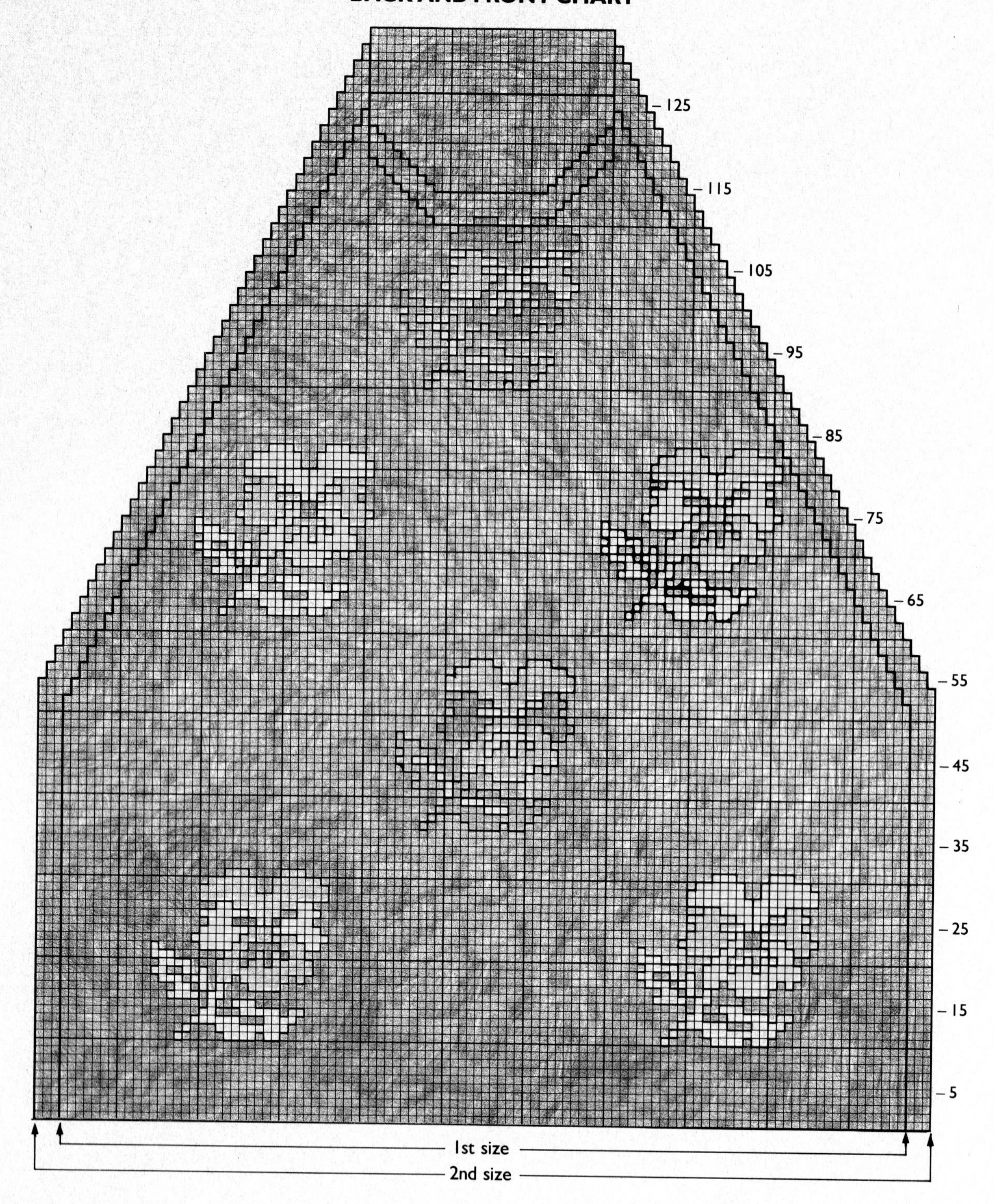

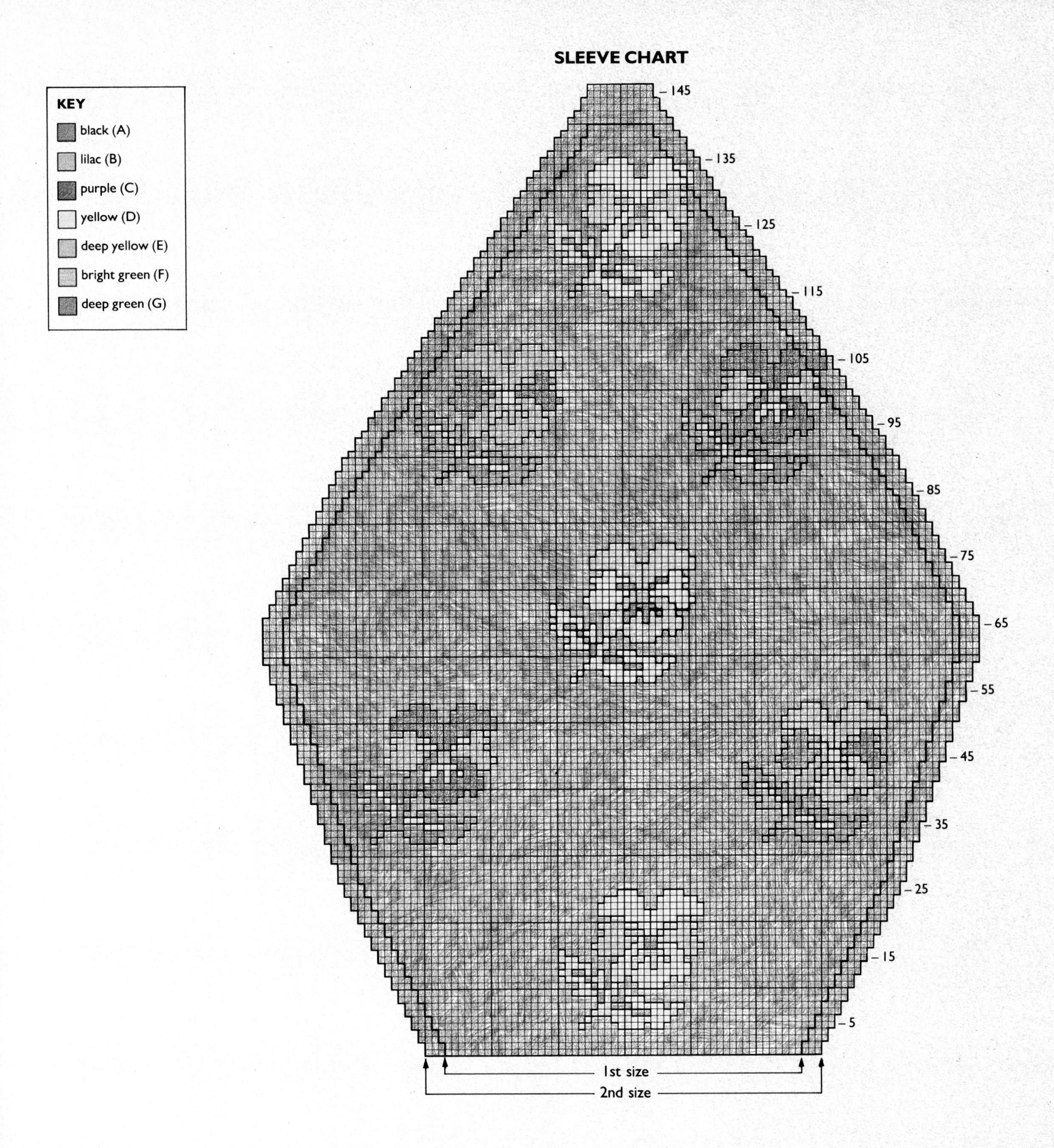
SLEEVE CHART
KEY
black (A)
lilac (B)
purple (C)
yellow (D)
deep yellow (E)
bright green (F)
deep green (G)
145
135
125
115
105
95
85
75
65
55
45
35
25
15
5
1st size
2nd size

With RS of work facing return to sts on spare needle, rejoin yarn, cast off centre 14 sts, patt to last 2 sts, work 2 tog.
Complete to match first side of neck.

SLEEVES

With 3¼mm (US 3) needles and A, cast on 36(40) sts.
Work in K1, P1 rib for 10cm/4in.
Increase row: Rib and inc 18(20) sts evenly across row – 54(60) sts.
Change to 4mm (US 5) needles. Beg with a K row cont in st st working from sleeve chart between appropriate lines for size required. Foll chart, inc 1 st at each end of 3rd and every foll alt row until there are 80(86) sts, then at each end of every foll 3rd row until there are 102(108) sts.
Cont without shaping, work 6 rows.

Shape raglan sleeve top

Foll chart, dec 1 st at each end of next and every foll alt row until 82(86) sts rem. Now dec 1 st at each end of next row, then at each end of foll alt row. Rep last 3 rows until 10 sts rem.
Cast off.

RAGLAN RIBS

4 alike (these are worked along the top shaped edges of the sleeves).
With 3¼mm (US 3) needles and A, K up 56(60) sts evenly along one side of shaped raglan sleeve top and work in K1, P1 rib for 3cm/1¼in.
Cast off loosely in rib.

COLLAR

Join raglan seams.
With RS of work facing, 3¼mm (US 3) circular needle and A, beg at centre of cast-off sts at centre front, K up 20(22) sts up right side of neck, 6 sts along top of raglan rib, 6(8) sts on top of sleeve, 6 sts along top of raglan rib, 28(30) sts across back neck, 6 sts along top of raglan rib, 6(8) sts on top of sleeve, 6 sts along top of raglan rib and 20(22) sts down left side of neck to centre front – 104(114) sts.
Mark beg of round with a coloured thread. Work 4 rounds K1, P1 rib.
Now cont to work in rows to form collar. Work in K1, P1 rib for 2·5cm/1in.
Change to 4mm (US 5) circular needle and cont in rows until work measures 10cm/4in. Cast off loosely in rib.

TO MAKE UP

Press lightly on wrong side of work with a steam iron or press under a damp cloth, avoiding ribbings.
Join side and sleeve seams.

TEXTURES

Simple but effective shapes have been chosen for these sweaters to show off subtle combinations of different yarns and stitch patterns.

The highly wearable **Large Check Cardigan** is knitted in soft tones of beige and cream used with a subtle fleck yarn. For the cool **Long Cable Vest** there is the original detail of a small half-belt at the back. Evenings are catered for with the pretty **Lace-stitch Sweater**, decorated with pearls for extra glamour. The classic good looks of the **Cable Sweater** would be effective in almost any colour, and to complete the section there is the delightful **Honeycomb Cardigan** knitted using an easy slip stitch in soft pastel colours with exciting textured yarns.

LARGE CHECK CARDIGAN

MEASUREMENTS

Two sizes	1st size	2nd size
To fit bust	81-86cm/32-34in	91-97cm/36-38in
Actual bust measurement	107cm/42in	118cm/46½in
Length	58.5cm/23in	61cm/24in
Sleeve seam	41cm/16in	43cm/17in

MATERIALS

Yarn

Any **double-knit** and **chunky** weight cotton yarn can be used as long as it knits up to the given tension.
150(175)g/6(7)oz double-knit beige cotton (A)
375(400)g/14(15) oz chunky fleck cotton (B)
400(450)g/15(16)oz chunky cream cotton (C)

Needles and other materials

1 pair of 4½mm (US 6) and 6mm (US 9) needles
6 large buttons

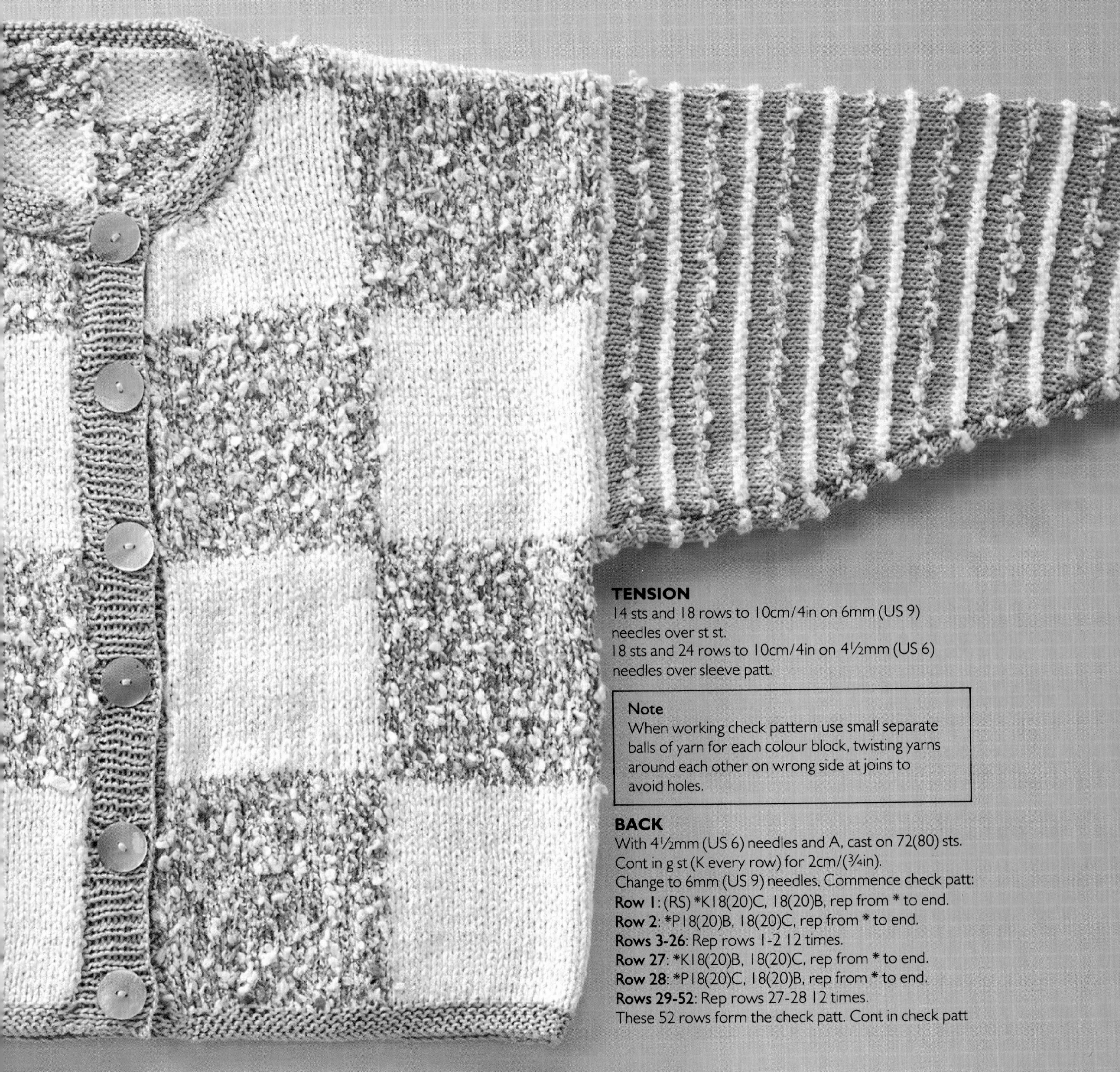

TENSION

14 sts and 18 rows to 10cm/4in on 6mm (US 9) needles over st st.

18 sts and 24 rows to 10cm/4in on 4½mm (US 6) needles over sleeve patt.

Note

When working check pattern use small separate balls of yarn for each colour block, twisting yarns around each other on wrong side at joins to avoid holes.

BACK

With 4½mm (US 6) needles and A, cast on 72(80) sts.
Cont in g st (K every row) for 2cm/(¾in).
Change to 6mm (US 9) needles. Commence check patt:
Row 1: (RS) *K18(20)C, 18(20)B, rep from * to end.
Row 2: *P18(20)B, 18(20)C, rep from * to end.
Rows 3-26: Rep rows 1-2 12 times.
Row 27: *K18(20)B, 18(20)C, rep from * to end.
Row 28: *P18(20)C, 18(20)B, rep from * to end.
Rows 29-52: Rep rows 27-28 12 times.
These 52 rows form the check patt. Cont in check patt

until work measures 58.5(61)cm/23(24)in from beg, ending with a WS row.

Shape shoulders
Cast off 8(9) sts at beg of next 6 rows. Cast off rem 24(26) sts.

LEFT FRONT

With 4½mm (US 6) needles and A, cast on 36(40) sts. Cont in g st for 2cm/¾in.
Change to 6mm (US 9) needles and cont in check patt as given for back until work measures 49(51)cm/19¼(20)in from beg, ending with a RS row.

Shape neck
Keeping patt correct, cast off 8 sts at beg of next row, then dec 1 st at neck edge on every alt row until 24(27) sts rem.
Cont without shaping until work matches back to shoulder shaping, ending with a WS row.

Shape shoulder
Cast off 8(9) sts at beg of next and foll alt row.
Work 1 row.
Cast off rem 8(9) sts.

RIGHT FRONT

Work as given for left front, reversing all shapings.

SLEEVES

Make 2. With 4½mm (US 6) needles and A, cast on 48 sts. Cont in g st for 2.5cm/1in.
Increase row: K and inc 12 sts evenly across row– 60 sts.

Cont in foll stripe patt, *at the same time*, inc 1 st at each end of 2nd and every foll 3rd row until there are 100 sts.
Row 1: (RS) With B, K.
Row 2: With B, K.
Rows 3-6: With A, beg with a K row work 4 rows st st.
Row 7: With C, K.
Row 8: With C, K.
Rows 9-12: Rep rows 3-6.
These 12 rows form the stripe patt.
Cont in stripe patt until work measures approx 41(43)cm/16(17)in from beg, ending with a patt row 2 or 8.
Cast off loosely with A.

NECKBAND

Join shoulder seams.
With RS of work facing 4½mm (US 6) needles and A, K up 24 sts up right front neck, 24(26) sts across back neck and 24 sts down left front neck–72(74) sts.
Cont in g st for 5cm/2in.
Cast off loosely.

BUTTONHOLE BAND

Fold neckband in half onto RS and slip stitch down.
With RS of work facing, 4½mm (US 6) needles and A, K up 92(96) sts evenly up right front edge and neckband.
Work 3 rows K1, P1 rib.
1st buttonhole row: (RS) Rib 5(4), *cast off 3, rib 13(14) sts including st used to cast off, rep from * 4 times more, cast off 3, rib to end.
2nd buttonhole row: Rib to end, casting on 3 sts over those cast off in previous row.
Work 3 rows rib.
Cast off loosely in rib.

BUTTON BAND

With RS of work facing, 4½mm (US 6) needles and A, K up 92(96) sts evenly down left front edge and neckband.
Work 8 rows K1, P1 rib. Cast off loosely in rib.

TO MAKE UP

Placing centre of cast-off edge of sleeves to shoulder seams, set in sleeves. Join side and sleeve seams. Sew on buttons.

LONG CABLE VEST

MEASUREMENTS

Three sizes	1st size	2nd size	3rd size
To fit bust	81-86cm/32-34in	86-91cm/34-36in	91-97cm/36-38in
Actual bust measurement	91cm/36in	99cm/39in	106cm/41¾in
Length	71cm/28in	73.5cm/29in	76.5cm/30¼in

MATERIALS

Yarn

Any **double-knit** weight cotton yarn can be used as long as it knits up to the given tension.
475(500:525)g/17(18:19)oz of chosen colour

Needles and other materials

1 pair each 3¾mm (US 4) and 4½mm (US 6) needles
1 cable needle
5 pearl buttons

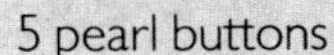

TENSION

22 sts and 24 rows to 10cm/4in on 4½mm (US 6) needles over reverse st st.

BACK

**With 3¾mm (US 4) needles, cast on 82(90:98) sts.
Cont in K1, P1 rib for 8cm/3in.
Increase row: Rib and inc 20 sts evenly across row – 102(110:118) sts.**
Change to 4½mm (US 6) needles and beg with a P row cont in reverse st st until work measures 45·5(48:51)cm/18(19:20)in from beg, ending with a K row.

Shape armholes

Cast off 10 sts at beg of next 2 rows. Dec 1 st at beg of next 10 rows – 72(80:88) sts.
Now cont without further shaping until work measures 63·5(66:69)cm/25(26:27¼) in from beg, ending with a K row.

Shape back neck

Next row: P19(23:27) sts, leave these sts on a spare needle, cast off centre 34 sts, P to end.
Complete left side of neck first keeping right side sts on spare needle.
Dec 1 st at neck edge on every foll alt row until 16 (16:24) sts rem.
3rd size only
Cast off 2 sts at beg of every foll alt row until 16 sts rem.
All sizes
Cont without shaping until work measures 71(73.5: 76·5)cm/28(29:30¼)in from beg, ending with a K row.
Cast off.
With WS of work facing return to sts on spare needle, rejoin yarn, K to end.
Complete to match first side of neck, reversing shapings.

FRONT

Work as given for back from ** to **.
Change to 4½mm (US 6) needles and commence patt:
Row 1: (RS facing) P15(19:23), K8, P56, K8, P15(19:23).
Row 2: K15(19:23), P8, K56, P8, K15(19:23).
Rows 3-6: Rep rows 1-2 twice more.
Row 7: P15(19:23), *sl next 4 sts onto cable needle and hold at back of work, K4, then K4 from cable needle, *P56, rep from * to *, P15(19:23).
Row 8: As row 2.
These 8 rows form the patt. Cont in patt until work measures 35.5(38:41)cm/14(15:16)in from beg, ending with a WS row.

Divide for front opening

Next row: Patt 47(51:55), K1, (P1, K1) 3 times, patt to end.
Next row: Patt 48(52:56), P1, (K1, P1) 3 times and turn, leaving rem sts on a spare needle.
Complete right side of neck first.
Work 8 rows straight as set, with 7 sts in rib at front edge.
1st buttonhole row: (RS facing) Rib 3, cast off 2, patt to end.
2nd buttonhole row: Patt to end, casting on 2 sts over those cast off in previous row.
Now cont as set until work matches back to armhole shaping, ending with a RS row.

Shape armhole

Cast off 10 sts at beg of next row, then dec 1 st at armhole edge on every alt row until 40(44:48) sts rem, *at the same time* when work measures 6cm/2½in from base of previous buttonhole work 1st and 2nd buttonhole rows again.
Now cont without shaping until work measures 53(55.5:58,5)cm/21(21¾:23)in from beg, ending with a WS row.

Shape neck

Cast off 16 sts at beg of next row, then dec 1 st at neck edge on every alt row until 16 sts rem.
Cont without further shaping until work matches back to shoulder.
Cast off.
With WS of work facing, return to sts on spare needle, cast on 8 sts, work (P1, K1) 4 times across these 8 sts, then patt across 47(51:55) sts on spare needle – 55(59:63) sts.
Next row: Patt 48(52:56), K1, (P1, K1) 3 times.
Complete to match first side of neck, omitting buttonholes and reversing all shapings.

NECKBAND

Join shoulder seams.
With RS of work facing and 3¾mm (US 4) needles, K up 7 sts from top of buttonhole band, 37 sts up right side of neck, 42 sts across back neck, 37 sts down left side of neck, then 7 sts from top of buttonband – 130 sts.
Work 2 rows K1, P1 rib.
1st buttonhole row: (WS facing) Rib to last 5 sts, cast off 2, rib to end.
2nd buttonhole row: Rib to end, casting on 2 sts over those cast off in previous row.
Work 6 rows rib, then rep 1st and 2nd buttonhole rows again.
Work 2 rows rib. Cast off loosely in rib.

ARMBANDS

Alike. With RS of work facing and 3¾mm (US 4) needles, K up 110 sts evenly around armhole edge.
Work 5 rows K1, P1 rib. Cast off in rib.

BELT

With 3¾mm (US 4) needles, cast on 12 sts.
Work in K1, P1 rib for 23cm/9in.
Cast off in rib.

TO MAKE UP

Carefully press pieces avoiding ribbing and cables using a steam iron or press under a damp cloth.
Stitch base of button band in place behind buttonhole band at centre front. Fold neckband in half onto RS and slip stitch in place.
Oversew double buttonhole. Position belt in centre of back 23cm/9in above cast-on edge and attach by sewing a button 2.5cm/1in from each end through both thicknesses. Join side seams. Sew buttons to front.

LACE-STITCH SWEATER

MEASUREMENTS

Two sizes	1st size	2nd size
To fit bust	86-91cm/34-36in	97-102cm/38-40in
Actual bust measurement	110cm/43¼in	120cm/47¼in
Length	61cm/24in	64cm/25¼in
Sleeve seam	43cm/17in	45.5cm/18in

MATERIALS

Yarn

Any **double-knit** weight cotton yarn can be used as long as it knits up to the given tension.
725(750)g/26(27)oz

Needles and other materials

1 pair each 3¼mm (US 3) and 4mm (US 5) needles
approx 450(500) small pearl beads

TENSION

22 sts and 30 rows to 10cm/4in on 4mm (US 5) needles over patt.

BACK

**With 3¼mm (US 3) needles, cast on 100(108) sts.
Cont in K2, P2 rib for 10cm/4in.
Increase row: Rib and inc 21(24) sts evenly across row – 121(132) sts.
Change to 4mm (US 5) needles and patt.
Row 1: (RS facing) *P1, K2 tog, yfwd, K5, yfwd, K2 tog tbl, P1, rep from * to end.
Row 2 and every foll alt row: *K1, P9, K1, rep from * to end.
Row 3: *P1, K1, yfwd, K2 tog tbl, K3, K2 tog, yfwd, K1, P1, rep from * to end.
Row 5: *P1, K2, yfwd, K2 tog tbl, K1, K2 tog, yfwd, K2, P1, rep from * to end.
Row 7: *P1, K3, yfwd, K3 tog, yfwd, K3, P1, rep from * to end.
Row 9: *P1, K2, K2 tog, yfwd, K1, yfwd, K2 tog tbl, K2, P1, rep from * to end.
Row 11: *P1, K1, K2 tog, yfwd, K3, yfwd, K2 tog tbl, K1, P1, rep from * to end.
Row 12: As row 2.
These 12 rows form the patt. Cont in patt until work measures 38(41)cm/15(16)in from beg, ending with a WS row.

Shape armholes

Keeping patt correct, cast off 11 sts at beg of next 2 rows – 99(110) sts **
Cont without shaping until work measures 61(64)cm/24(25¼)in from beg, ending with a WS row.

Shape shoulders

Cast off 10(12) sts at beg of next 6 rows. Cast off rem 39(38) sts.

FRONT

Work as given for back from ** to **.
Cont without shaping until work measures 53(56)cm/ 21(22)in from beg, ending with a WS row.

Shape neck

Next row: (RS facing) Patt 41(46) and turn, leaving rem sts on a spare needle.
Complete left side of neck first.
Dec 1 st at neck edge on every row until 30(36) sts rem.
Cont without further shaping until work matches back to shoulder shaping, ending at armhole edge.

Shape shoulder

Cast off 10(12) sts at beg of next and foll alt row. Work 1 row.
Cast off rem 10(12) sts.
With RS of work facing return to sts on spare needle, rejoin yarn, cast off centre 17(18) sts, patt to end.
Complete as given for first side of neck.

SLEEVES

Make 2. With 3¼mm (US 3) needles, cast on 56 sts.
Cont in K2, P2 rib for 10cm/4in.
Increase row: Rib and inc 54 sts evenly across row – 110 sts.
Change to 4mm (US 5) needles and commence patt as given for back.
Cont in patt until work measures 43(45.5)cm/17(18)in from beg, ending with a WS row.

Shape top

Keeping patt correct, cast off 11 sts at beg of next 2 rows. Dec 1 st at beg of foll 40 rows. Cast off rem 48 sts.

COLLAR

Join right shoulder seam.
With RS of work facing and 3¼mm (US 3) needles K up 27 sts down left side of neck, 16 sts from centre front neck, 27 sts up right side of neck and 38 sts from back neck – 108 sts.
Cont in K2, P2 rib for 2.5cm/1in, ending with a WS row.
Increase row: *(work K1, P1 into next st) twice, P2, rep from * to end – 162 sts.
Change to 4mm (US 5) needles.
Next row: *K2, P4, rep from * to end.
Next row: *K4, P2, rep from * to end.
These 2 rows form the rib patt. Cont in rib patt until work measures 10 cm/4in from beg. Cast off loosely in rib.

TO MAKE UP

Join left shoulder and collar seam.
Set in sleeves gathering fullness over shoulder.
Join side and sleeve seams.
Sew a pearl bead to the centre of each lace diamond as shown in the photograph.

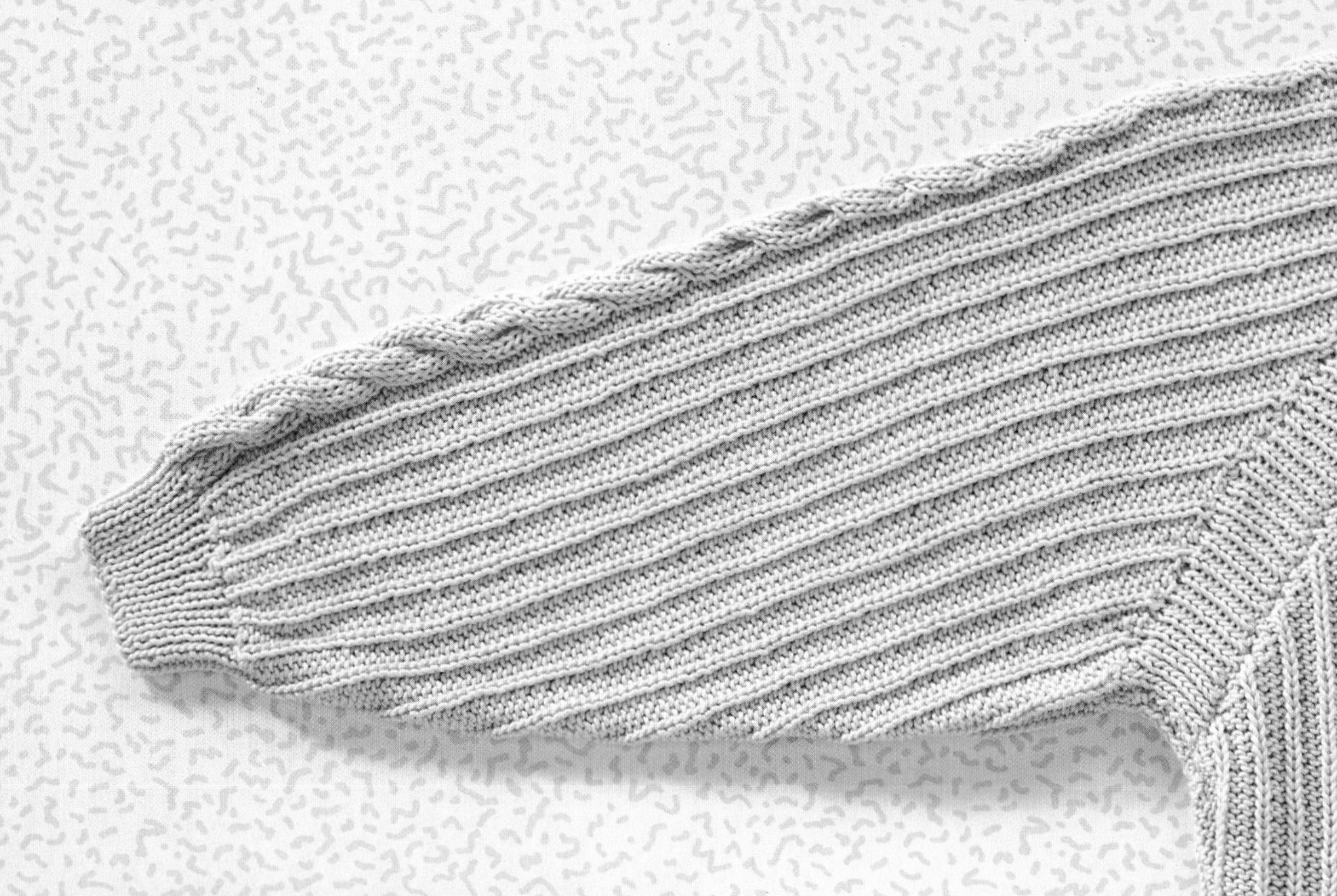

CABLE SWEATER

MEASUREMENTS

Three sizes	1st size	2nd size	3rd size
To fit bust	81-86cm/32-34in	86-91cm/34-36in	91-97cm/36-38in
Actual bust measurement	106cm/41¾in	113cm/44½in	120cm/47¼in
Length to back neck	54cm/21¼in	57cm/22½in	60.5cm/23¾in
Sleeve seam	35.5cm/14in	35.5cm/14in	35.5cm/14in

MATERIALS

Yarn

Any **double-knit** weight cotton yarn can be used as long as it knits up to the given tension.
825(850:875)g/29(30:31)oz of chosen colour

Needles

1 pair each 3¾mm (US 4) and 4½mm (US 6) needles
1 cable needle

TENSION

23 sts and 26 rows to 10cm/4in on 4½mm (US 6) needles over rib patt.
33 st cable panel measures 10cm/4in wide.

BACK

With 3¾mm (US 4) needles, cast on 104(112:120) sts.
Work in K1, P1 rib for 5cm/2in.

Increase row: Rib and inc 28 sts evenly across row – 132(140:148) sts.
Change to 4½mm (US 6) needles and commence patt.
Row 1: (RS) *P3, K1* rep from * to * 11(12:13) times more, **P3, K8**, rep from ** to ** twice more, now rep from * to * 12(13:14) times, P3.
Row 2: *K3, P1*, rep from * to * 11(12:13) times more, **K3, P8**, rep from ** to ** twice more, now rep from * to * 12(13:14) times, K3.
Rows 3-6: Rep rows 1-2 twice more.
Row 7: *P3, K1*, rep from * to * 11(12:13) times more, **P3, sl next 4 sts onto cable needle and hold at back of work, K4 then K4 from cable needle**, rep from ** to ** twice more, now rep from * to * 12(13:14) times, P3.
Row 8: As row 2.
These 8 rows form the patt. Cont in patt until work measures 28(30.5:33) cm/11(12:13) in from beg, ending with a WS row ***.

Shape raglan armholes
Rows 1-2: Patt and dec 1 st at each end of row.
Row 3: Patt to end.
Rep rows 1-3 21(22:23) times more.
1st and 2nd sizes only
Work rows 1-2(row 1) again.
All sizes
Cast off rem 40(46:52) sts.

FRONT

Work as given for back to ***.
Shape raglan armholes.
Rep rows 1-3 of back raglan shaping 16(17:18) times – 68(72:76) sts.
1st and 2nd sizes only
Work rows 1-2(row 1) again – 64(70) sts.

All sizes
Shape neck
Next row: (RS facing) Patt 21(work 2 tog, patt 19: work 2 tog, patt 20), turn leaving rem sts on a spare needle.
Complete left side of neck first.
Still dec at armhole edge as before, dec 1 st at neck edge on every foll alt row until 2 sts rem.
Next row: Work 2 tog and fasten off.
With RS of work facing, return to sts on spare needle, rejoin yarn, cast off centre 22(28:32) sts and complete to match first side.

SLEEVES

Make 2. With 3¾mm (US 4) needles, cast on 48 sts.
Work in K1, P1 rib for 5cm/2in.
Increase row: Rib and inc 24 sts evenly across row – 72 sts.
Change to 4½mm (US 6) needles and work in foll patt, *at the same time*, inc and work into rib patt 1 st at each end of 2nd and every foll 3rd row until there are 118 sts.

Row 1: (RS) *K1, P3, rep from * 7 times more, K8, **P3, K1, rep from ** 7 times more.
Row 2: *P1, K3, rep from * 7 times more, P8, **K3, P1, rep from ** 7 times more.
Rows 3-6: Rep rows 1-2 twice more.
Row 7: *K1, P3, rep from * 7 times more, sl next 4 sts onto cable needle and hold at back of work, K4 then K4 from cable needle, **P3, K1, rep from ** 7 times more.
Row 8: As row 2.
These 8 rows form the patt. Cont in patt without shaping until work measures 35·5cm/14in from beg, ending with a WS row.

Shape raglan sleeve top
Keeping patt correct, dec 1 st at each end of every row until 66(86:100) sts rem.
Rep rows 1-3 of back raglan shaping until 10(14:16) sts rem, ending with a WS row.
Cast off.

RAGLAN RIBS

4 alike (these are worked along the top shaped edges of the sleeves).
With 3¾mm (US 4) needles and RS of work facing, K up 68(72:76) sts evenly along one side of shaped raglan sleeve top and work in K1, P1 rib for 4cm/1½in.
Cast off loosely in rib.

NECKBAND

Join raglan seams leaving left back raglan open.
With RS of work facing and 3¾mm (US 4) needles K up 5 sts along top of left raglan rib, 8(12:14) sts on top of left sleeve, 5 sts along top of raglan rib, 14 sts down left side of neck, 18(24:28) sts across centre front neck, K up 14 sts up right side of neck, 5 sts along raglan rib, 8(12:14) sts on top of right sleeve, 5 sts along right raglan rib and 34(40:46) sts across back neck – 128(136:144) sts.
Work in K1, P1 rib for 8cm/3in.
Cast off loosely in rib.

TO MAKE UP

Join remaining raglan seam and neckband. Fold neckband in half onto RS and slip stitch in place. Join side and sleeve seams.

HONEYCOMB CARDIGAN

MEASUREMENTS

Two sizes	1st size	2nd size
To fit bust	81-91cm/32-36in	97-102cm/38-42in
Actual bust measurement	118cm/46½in	133cm/52½in
Length	58.5cm/23in	61cm/24in
Sleeve seam	43cm/17in	45.5cm/18in

MATERIALS

Yarn

Any **chunky-weight** cotton yarn can be used as long as it knits up to the given tension.
425(450)g/15(16)oz chunky fleck cotton (A)
100(125)g/4(5)oz chunky bouclé cotton (B)
75(100)g/3(4)oz chunky textured cotton (C)
75(75)g/3(3)oz chunky textured cotton (D)

Needles and other materials

1 pair each 4½mm (US 6) and 5½mm (US 8) needles
4½mm (US 6) circular needle
5 medium buttons

TENSION

16 sts and 20 rows to 10cm/4in on 5½mm (US 8) needles over st st.
14 sts and 24 rows to 10cm/4in on 5½mm (US 8) needles over patt.

BACK

With 4½mm (US 6) needles and A, cast on 94(100) sts.
Cont in K1, P1 rib for 2.5cm/1in.
Change to 5½mm (US 8) needles and beg with a K row cont in st st until work measures 58.5(61)cm/23(24)in from beg, ending with a P row.

Shape shoulders

Cast off 10(11) sts at beg of next 6 rows. Cast off rem 34 sts.

LEFT FRONT

With 4½mm (US 6) needles and A, cast on 40(48) sts.
Cont in K1, P1 rib for 2.5cm/1in.
Change to 5½mm (US 8) needles and commence patt.

Row 1: (RS) With B, K.
Row 2: With B, K.
Row 3: With A, K.
Row 4: With A, P.
Row 5: With B, K.
Row 6: With B, K1, *K2 wrapping yarn twice round needle for each st, K6, rep from * ending last rep K5.
Row 7: With C, K5, * sl 2 P-wise allowing extra loops to fall, K6, rep from * ending last rep K1.
Row 8: With C, P1, *sl 2 P-wise, P6, rep from * ending last rep P5.
Rows 9-12: Rep rows 7-8 twice more.
Rows 13-17: Rep rows 1-5.
Row 18: With B, K5, *K2 wrapping yarn twice round needle for each st, K6, rep from * ending last rep K1.
Row 19: With D, K1, * sl 2 P-wise allowing extra loops to fall, K6, rep from * ending last rep K5.
Row 20: With D, P5, * sl 2 P-wise allowing extra loops to fall, P6, rep from * ending last rep P1.
Rows 21-24: Rep rows 19-20 twice more.
These 24 rows form the patt. Cont in patt until work measures 33cm/13in from beg, ending at front edge.

Shape front neck
Keeping patt correct, dec 1 st at beg of next row and at neck edge on every foll 4th(3rd) row until 26(29) sts rem.
Cont without shaping until work matches back to shoulder shaping, ending at armhole edge.

Shape shoulder
Cast off 9(10) sts at beg of next and foll alt row. Work 1 row. Cast off rem 8(9) sts.

RIGHT FRONT

Work as given for left front reversing shaping.

SLEEVES

Make 2. With 4½mm (US 6) needles and A, cast on 36 sts. Work in K1, P1 rib for 2.5cm/1in.
Increase row: Rib and inc 18 sts evenly across row – 54 sts.
Change to 5½mm (US 8) needles and beg with a K row cont in st st, inc 1 st at each end of 2nd and every foll 3rd row until there are 80 sts. Cont without shaping until work measures 43(45.5)cm/17(18)in from beg, ending with a P row. Cast off loosely.

FRONT BAND

Join shoulder seams.
With 4½mm (US 6) circular needle, A and RS of work facing, K up 52 sts to beg of neck shaping, 36(40) sts up right front neck, 30 sts from across back neck and 36(40) sts down left front neck, then 52 sts to lower edge – 206(214) sts. Work in rows.
Work 3 rows K1, P1 rib.
1st buttonhole row: (RS facing) Rib 4, * cast off 2, rib 9 including st used to cast off, rep from * 3 times more, cast off 2, rib to end.
2nd buttonhole row: Rib to end, casting on 2 sts over those cast off in previous row.
Work 2 rows rib.
Cast off in rib.

TO MAKE UP

Placing centre of cast-off edge of sleeves to shoulder seams, set in sleeves.
Join side and sleeve seams.
Sew on buttons.

GEOMETRICS

There is a casual, sporty mood for this section featuring some bright and breezy designs.

The distinctive **Harlequin Jacket** has a mandarin collar and a bold pattern of broken stripes and diamonds against a white ground. Then follows the **Floral Block Sweater** with an interesting combination of plain cottons mixed with textured and multicoloured yarns to create a rich fabric. There's a bright **Vertical Striped Cardigan** with gathered sleeves, and a **Classic Striped Cardigan** – totally wearable and terribly chic with its broad stripes in cream and black. To end, the **Stripe and Block Sweater** with a stylish boat neck and wide boxy shape, ideal for sailing or the beach, is very easy to knit.

HARLEQUIN JACKET

MEASUREMENTS

Two sizes	1st size	2nd size
To fit bust	81-86cm/32-34in	91-97cm/36-38in
Actual bust measurement	110cm/43¼in	115cm/45¼in
Length	60cm/23¾in	61.5cm/24¼in
Sleeve seam	45cm/17¾in	50cm/19¾in

MATERIALS

Yarn

Any **double-knit** weight cotton yarn can be used as long as it knits up to the given tension.
600(650)g/21(23)oz white (A)
75(100)g/3(4)oz blue (B)
75(100)g/3(4)oz lime green (C)
125(150)g/5(6)oz orange (D)
75g/3oz red (E)
75g/3oz green (F)

Needles and other materials

1 pair each 3¼mm (US 3) and 4mm (US 5) needles
6 square buttons – 3 red, 3 green

TENSION

21 sts and 24 rows to 10cm/4in on 4mm (US 5) needles over st st.

Note

- When reading charts, work K rows (odd-numbered rows) from right to left and P rows (even-numbered rows) from left to right.
- When working the motifs, wind off small amounts of the required colours so that each motif can be worked separately, twisting yarns around each other on wrong side at joins to avoid holes. On the 2-colour horizontal bands yarn can be carried on wrong side of work over not more than 3 stitches at a time to keep fabric elastic.

FRONTS CHART

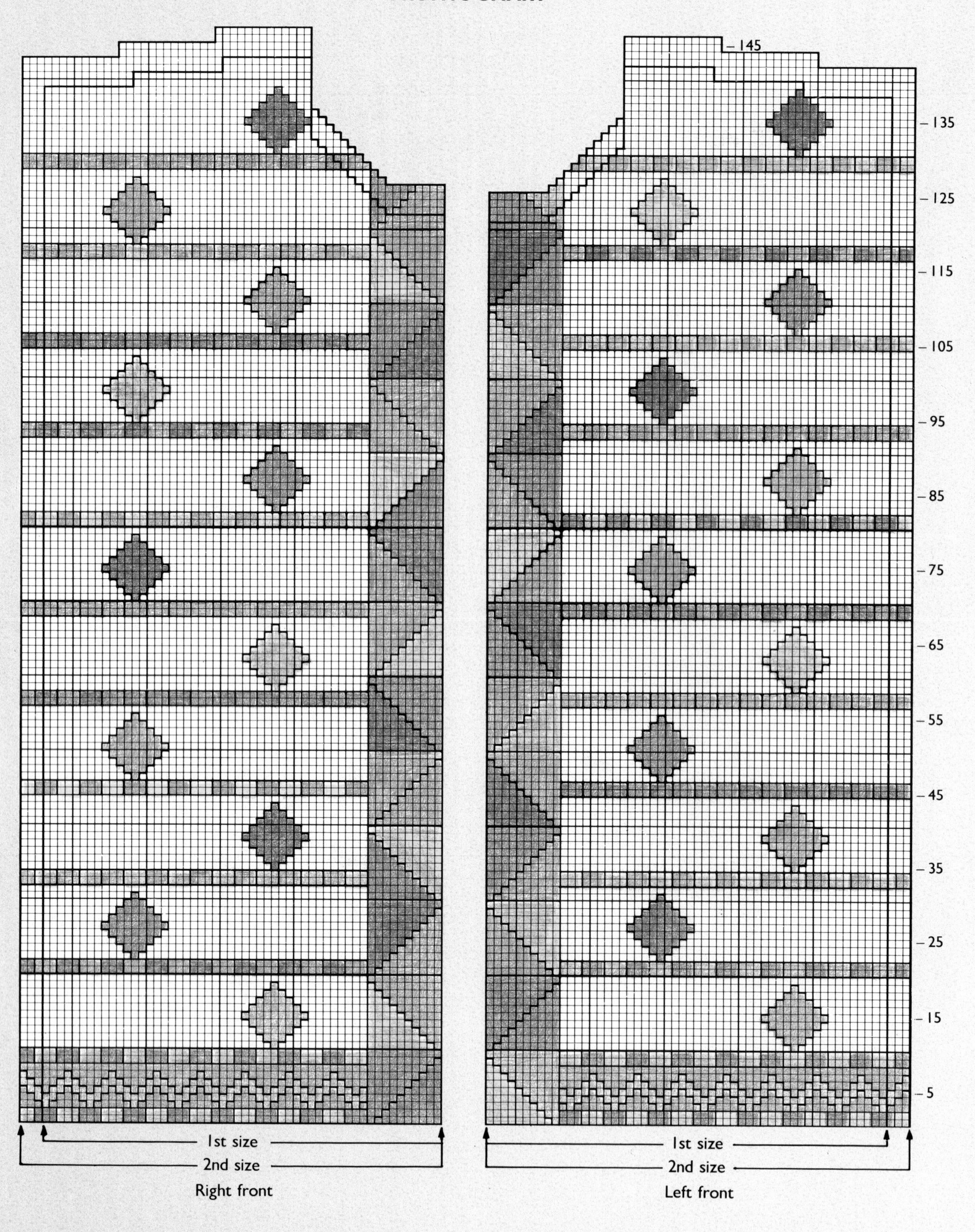

BACK AND SLEEVE CHART

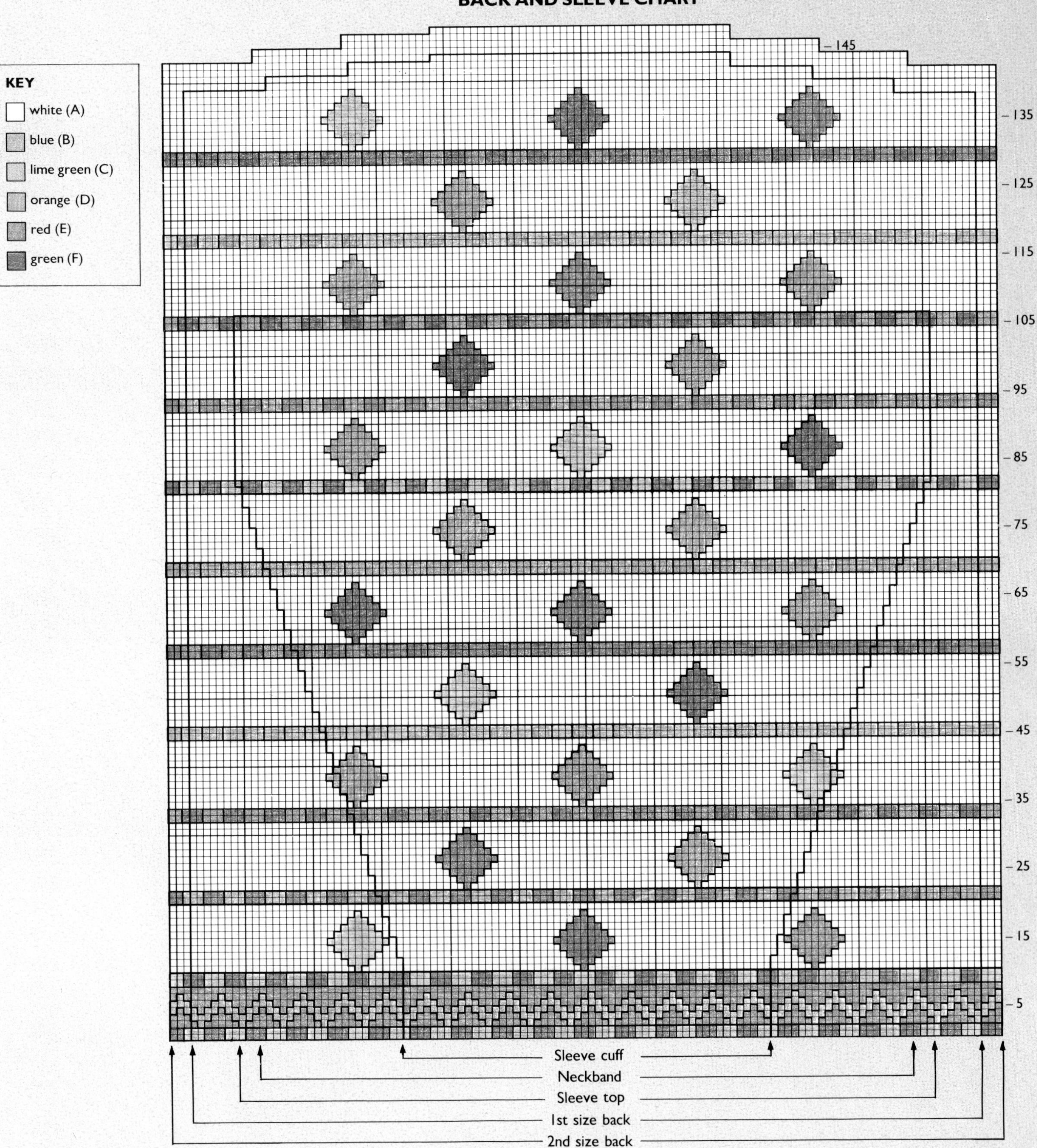

BACK

With 3¼mm (US 3) needles and A, cast on 116(122) sts. Cont in K1, P1 rib for 2·5cm/1in.
Change to 4mm (US 5) needles and beg with a K row cont in st st working from back chart between appropriate lines for size required.
Work 138(142) rows.

Shape shoulders

Follow chart, cast off 12(13) sts at beg of next 6 rows. Leave rem 44 sts on a spare needle.

LEFT FRONT

With 3¼mm (US 3) needles and A, cast on 54(56) sts. Cont in K1, P1 rib for 2·5cm/1in, inc 1 st at end of last row on 2nd size only – 54(57) sts.
Change to 4mm (US 5) needles and beg with a K row cont in st st working from left front chart.
Work 121(125) rows.

Shape neck

Next row: (WS facing – foll chart) Cast off 8 sts, patt to end.
Dec 1 st at neck edge on every row until 36(39) sts rem.
Cont without shaping until row 138(142) of chart has been completed.

Shape shoulder

Cast off 12(13) sts at beg of next and foll alt row. Work 1 row.
Cast off rem 12(13) sts.

RIGHT FRONT

Work as given for left front, working from appropriate chart and reversing all shapings by working 1 more row before neck and shoulder shaping.

SLEEVES

Make 2. With 3¼mm (US 3) needles and A, cast on 54 sts. Cont in K1, P1 rib for 2·5cm/1in.
Change to 4mm (US 5) needles and beg with a K row cont in st st working from back chart between appropriate lines for sleeve cuff. Foll chart, inc 1 st at each end of 13th and every foll 3rd row until there are 102 sts.
Cont without shaping until row 92(104) of chart has been completed.
Now work rows 1-10 from chart again, reading between markers for sleeve top.
Cast off loosely.

BUTTONHOLE BAND

With RS of work facing, 3¼mm (US 3) needles and A, K up 110(114) sts evenly up right front edge.
Work 3 rows K1, P1 rib.
1st buttonhole row: Rib 4, *cast off 2, rib 18(19) sts including st used to cast off, rep from * 4 times more, cast off 2, rib to end.
2nd buttonhole row: Rib to end, casting on 2 sts over those cast off in previous row.
Work 2 rows rib.
Cast off in rib.

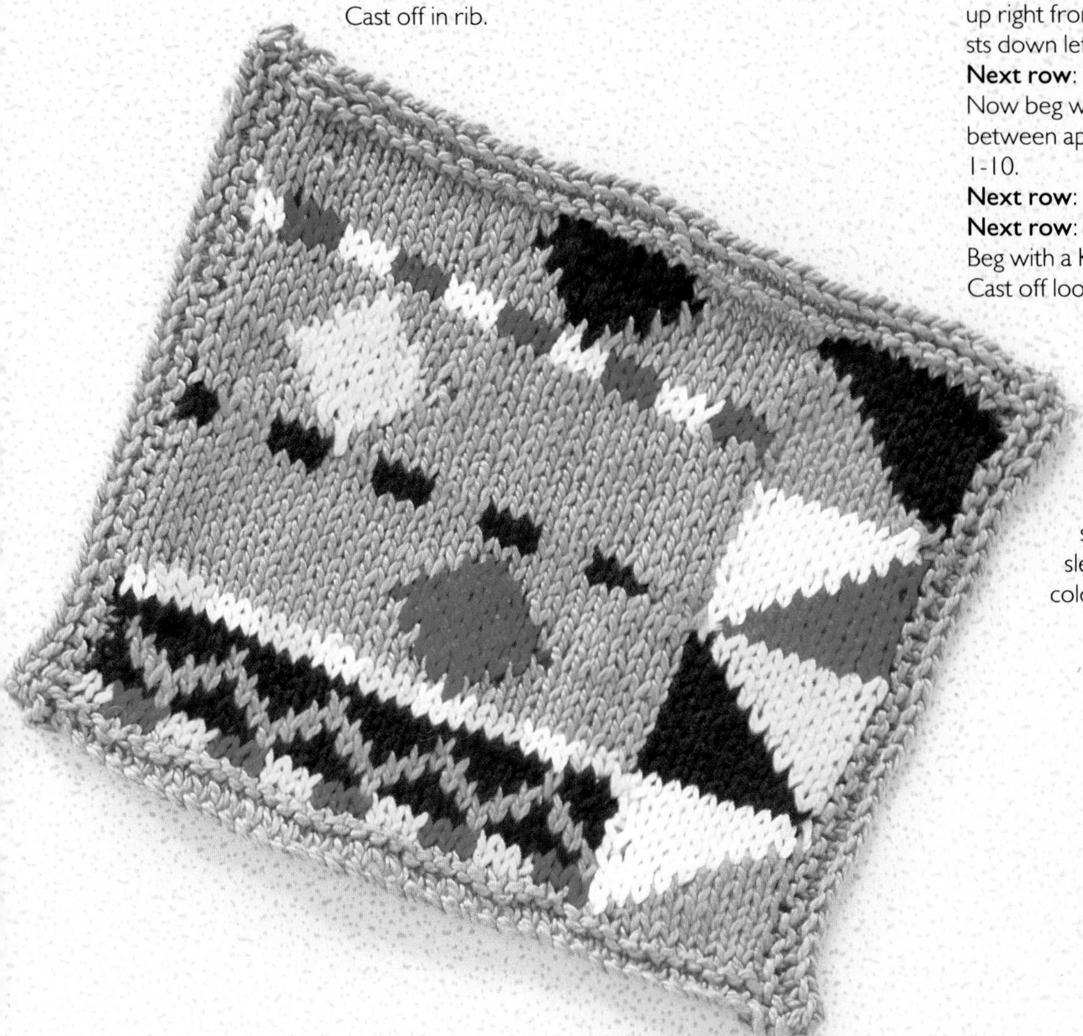

BUTTON BAND

With RS of work facing, 3¼mm (US 3) needles and A, K up 110(114) sts evenly down left front edge.
Work 7 rows K1, P1 rib.
Cast off in rib.

NECKBAND

Join shoulder seams.
With RS of work facing, 3¼mm (US 3) needles and A, starting at inner edge of buttonhole band, K up 26 sts up right front neck, K across 44 back neck sts, K up 26 sts down left side of neck – 96 sts.
Next row: P.
Now beg with a K row work in st st from back chart between appropriate lines for neckband. Work rows 1-10.
Next row: With A, K.
Next row: (fold-line) With A, K.
Beg with a K row, cont in A and st st, work 12 rows.
Cast off loosely.

TO MAKE UP

Fold neckband in half onto RS and sew short front neck edges together.
Turn RS out and slip stitch cast-off edge in place.
Placing centre of cast-off edge of sleeves to shoulder seams, set in sleeves. Join side and sleeve seams. Sew on buttons alternating colours as shown.

FLOWER BLOCK SWEATER

MEASUREMENTS

Three sizes	1st size	2nd size	3rd size
To fit bust	86-91cm/34-36in	91-97cm/36-38in	97-102cm/38-40in
Actual bust measurement	99.5cm/39½in	107cm/42in	115cm/45¼in
Length	58cm/22¾in	60cm/23½in	63cm/24¾in
Sleeve seam	41cm/16in	43cm/17in	45.5cm/18in

MATERIALS

Yarn

Any **double-knit** weight cotton blends can be used as long as they knit up to the given tension.

300(325:350)g/11(12:13)oz dk cotton blue (A)
75(100:125)g/3(4:5)oz dk cotton fleck (B)
50(75:75)g/2(3:3)oz dk cotton viscose brown (C)
50(75:75)g/2(3:3)oz dk cotton turquoise (D)
50(75:75)g/2(3:3)oz dk cotton cream (E)
75(100:100)g/3(4:4)oz dk cotton black (F)
100(125:125)g/4(5:5)oz cotton bouclé (G)

Needles

1 pair each 3¾mm (US 4) and 4½mm (US 6) needles

TENSION

20½ sts and 27 rows to 10cm/4in on 4½mm (US 6) needles over patt.

Note

- When reading charts, work K rows (odd-numbered rows) from right to left and P rows (even-numbered rows) from left to right.
- When working the floral motif and blocks, wind off small amounts of the required colours so that each colour area can be worked separately, twisting yarns around each other on wrong side at joins to avoid holes.
- On the 2-colour horizontal bands yarn can be carried on wrong side of work over not more than 3 stitches at a time to keep fabric elastic.

BACK

**With 3¾mm (US 4) needles and A, cast on 78(86:94) sts. Cont in K2, P2 rib.

Row 1: (RS) K2, *P2, K2, rep from * to end.

Row 2: P2, *K2, P2, rep from * to end.

Rep these 2 rows for 10cm/4in ending with a RS row.

Increase row: Rib and inc 24 sts evenly across row – 102(110:118) sts.

Change to 4½mm (US 6) needles. Beg with a K row, cont in st st working from back chart between appropriate lines for size required.**

Work 130(136:142) rows.

Shape shoulders

Foll chart, cast off 8(9:10) sts at beg of next 8 rows.
Cast off rem 38 sts.

FRONT

Work as given for back from ** to **
Work 110(116:122) rows.

Shape neck

Next row: (RS facing – foll chart) Patt 41(45:49) sts and turn, leaving rem sts on a spare needle.

Complete left side of neck first. Dec 1 st at neck edge on every foll alt row until 32(36:40) sts rem.

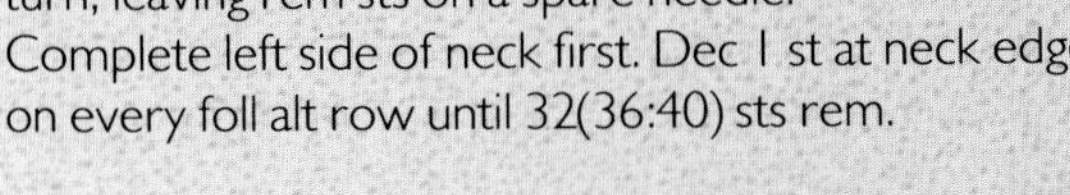

BACK AND FRONT CHART

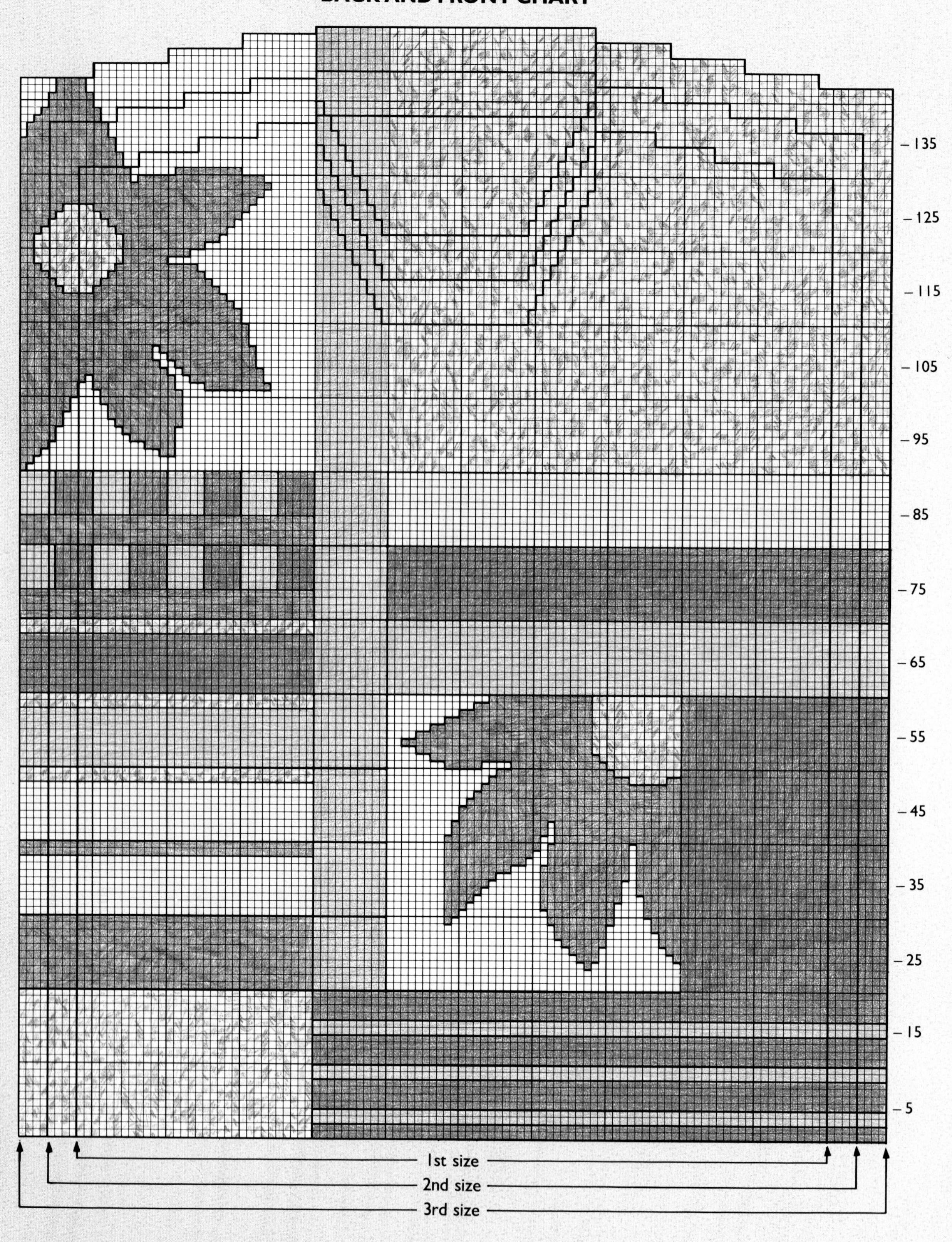

Cont without shaping until work matches back to shoulder shaping, ending at armhole edge.

Shape shoulder

Cast off 8(9:10) sts at beg of next and 2 foll alt rows.
Work 1 row.
Cast off rem 8(9:10) sts.
With RS of work facing return to sts on spare needle, rejoin yarn, cast off centre 20 sts, patt to end.
Complete as given for first side of neck.

SLEEVES

Make 2. With 3¾mm (US 4) needles and A, cast on 48 sts. Cont in K2, P2 rib for 10cm/4in.
Increase row: Rib and inc 16 sts evenly across row. 64 sts.
Change to 4½mm (US 6) needles. Beg with a K row, cont in st st working from sleeve chart, *at the same time*, inc 1 st at each end of every 4th row until there are 100 sts.
Cont without shaping until row 84(90:96) of chart has been completed.
Cast off loosely.

NECKBAND

Join right shoulder seam.
With RS of work facing, 3¾mm (US 4) needles and A, K up 22 sts down left side of neck, 18 sts from centre front, 22 sts up right side of neck and 36 sts across back neck – 98 sts.
Starting with row 1, work in K2, P2 rib as given for back for 8cm/3in. Cast off loosely in rib.

TO MAKE UP

Press carefully avoiding ribbing using a steam iron or press under a damp cloth.
Join left shoulder and neckband seam. Fold neckband in half onto RS and slip stitch in place. Placing centre of cast-off edge of sleeve to shoulder seam, set in sleeves. Join side and sleeve seams.

SLEEVE CHART

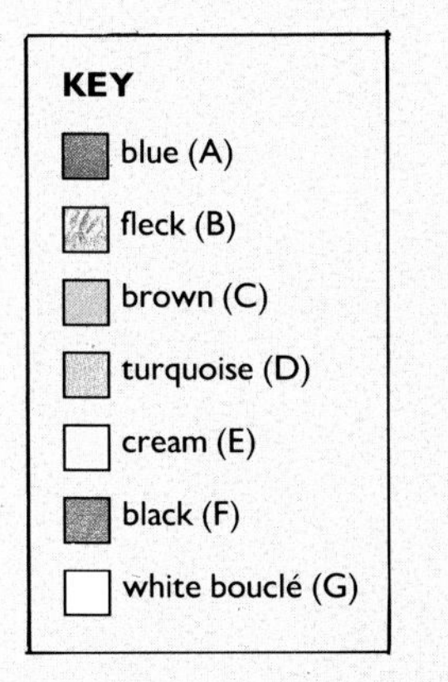

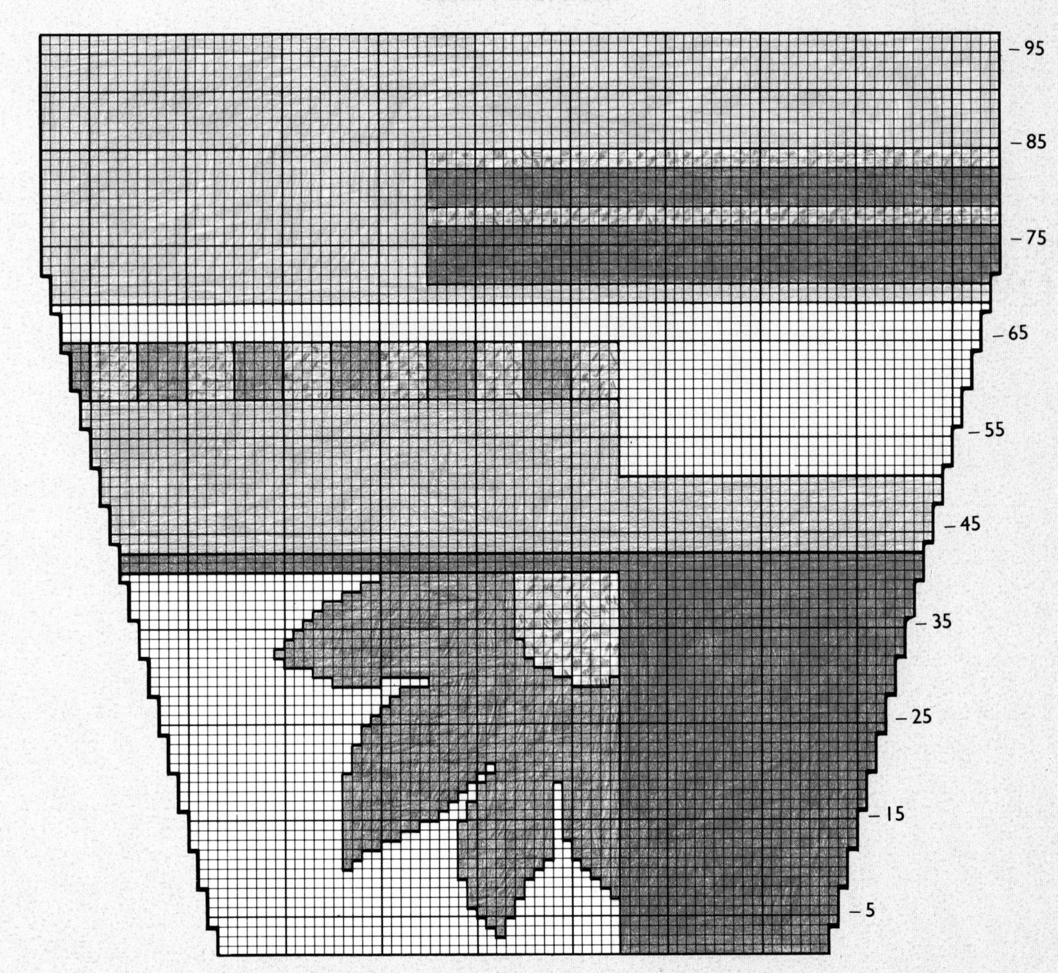

VERTICAL STRIPE CARDIGAN

MEASUREMENTS

Three sizes	1st size	2nd size	3rd size
To fit bust	86-91cm/34-36in	91-97cm/36-38in	97-102cm/38-40in
Actual bust measurement	97cm/38in	102.5cm/40¼in	108cm/42½in
Length	58.5cm/23in	61cm/24in	64cm/25¼in
Sleeve seam	43cm/17in	45.5cm/18in	48cm/19in

MATERIALS

Yarn

Any **double-knit** weight cotton yarn can be used as long as it knits up to the given tension.
150(175:200)g/6(7:8)oz white (A)
150(175:175)g/6(7:7)oz blue (B)
150(150:150)g/6(6:6)oz pink (C)
175(175:200)g/7(7:8)oz yellow (D)

Needles and other materials

1 pair each 3¼mm (US 3) and 4mm (US 5) needles
3¼mm (US 3) circular needle
8 buttons

TENSION

22 sts and 27 rows to 10cm/4in on 4mm (US 5) needles over st st.

> **Note**
> When working vertical stripes use small separate balls of A and twist yarns on wrong side of work when changing colour to avoid a hole.

BACK

With 3¼mm (US 3) needles and A, cast on 90(96:102) sts. Cont in K1, P1 rib for 10cm/4in.
Increase row: Rib and inc 16 sts evenly across row – 106(112:118) sts.
Change to 4mm (US 5) needles and commence vertical stripe patt.
Row 1: (RS) K3(6:9)A, 31D, 3A, 32C, 3A, 31B, 3(6:9)A.
Row 2: P3(6:9)A, 31B, 3A, 32C, 3A, 31D, 3(6:9)A.
These 2 rows form the stripe patt. Cont in patt until

work measures 35.5(38:41)cm/14(15:16)in from beg, ending with a WS row.

Shape armholes
Keeping patt correct, cast off 8 sts at beg of next 2 rows. Dec 1 st at beg of foll 8 rows – 82(88:94) sts. Cont without further shaping until work measures 58.5 (61:64)cm/23(24:25¼)in from beg, ending with a WS row.

Shape shoulders
Cast off 8(9:10) sts at beg of next 6 rows. Leave rem 34 sts on a spare needle.

LEFT FRONT

With 3¼mm (US 3) needles and A, cast on 42(44:46) sts. Cont in K1, P1 rib for 10cm/4in.
Increase row: Rib and inc 9(10:11) sts evenly across row – 51(54:57) sts.
Change to 4mm (US 5) needles and commence vertical stripe patt.
Row 1: (RS) K3A, 13(14:15)D, 3A, 13(14:15)C, 3A, 13(14:15)B, 3A.
Row 2: P3A, 13(14:15)B, 3A, 13(14:15)C, 3A, 13(14:15)D, 3A.
These 2 rows form the stripe patt. Cont in patt until work measures same as back to armhole shaping, ending at side edge.

Shape armhole
Cast off 8 sts at beg of next row. Dec 1 st at armhole edge on every foll alt row until 39(42:45) sts rem. Cont without further shaping until work measures 51.5(53:56)cm/20¼(21:22)in from beg, ending at front edge.

Shape neck
Cast off 10 sts at beg of next row. Dec 1 st at neck edge on every foll alt row until 24(27:30) sts rem. Cont without further shaping until work matches back to shoulder shaping, ending at armhole edge.

Shape shoulder
Cast off 8(9:10) sts at beg of next and foll alt row. Work 1 row. Cast off rem 8(9:10) sts.

RIGHT FRONT

Work as given for left front, reversing all shapings.

RIGHT SLEEVE

With 3¼mm (US 3) needles and A, cast on 48 sts.
Cont in K1, P1 rib for 10cm/4in.
Increase row: Rib and inc 24 sts evenly across row – 72 sts.
Change to 4mm (US 5) needles and beg with a P row cont in st st work 1 row A.
Now cont in stripe sequence of 16 rows B, 4 rows A, 16 rows C, 4 rows A, 16 rows D, 4 rows A, *at the same time*, inc 1 st at each end of 3rd and every foll 4th row until there are 96 sts.
Keeping stripe sequence correct, cont without shaping until work measures 43(45.5:48)cm/17(18:19)in from beg, ending with a P row.

Shape top

Cast off 8 sts at beg of next 2 rows. Dec 1 st at beg of every row until 50 sts rem. Cast off.

LEFT SLEEVE

Work as given for right sleeve but work in stripe sequence of 16 rows D, 4 rows A, 16 rows C, 4 rows A, 16 rows B, 4 rows A.

NECKBAND

Join shoulder seams.
With RS of work facing, 3¼mm (US 3) needles and A, K up 26 sts up right side of neck, K across 34 sts on back neck and K up 26 sts down left side of neck – 86 sts.
Cont in K1, P1 rib for 7cm/2¾in. Cast off loosely in rib.

BUTTONHOLE BAND

Fold neckband in half onto RS and slip stitch in place.
With RS of work facing, circular needle 3¼mm (US 3) and A, K up 126(132:138) sts evenly up right front edge and neckband. Work in rows.
Work 3 rows K1, P1 rib.
1st buttonhole row: (RS facing) Rib 4, *cast off 3 sts, rib 7(9:11) including st used to cast off*, rep from * to *, rib 1, **cast off 3 sts, rib 18 including st used to cast off**, rep from ** to ** 3 times more, rep from * to * again, cast off 3 sts, rib to end.
2nd buttonhole row: Rib to end, casting on 3 sts over those cast off in previous row.
Work 3 rows rib. Cast off in rib.

BUTTON BAND

With RS of work facing, 3¼mm (US 3) circular needle and A, K up 126(132:138) sts evenly down neckband and left front edge. Work in rows.
Work 8 rows K1, P1 rib. Cast off in rib.

TO MAKE UP

Press lightly on wrong side of work avoiding ribbing using a steam iron or press under a damp cloth.
Set in sleeves, gathering top to fit. Join side and sleeve seams.
Sew on buttons.

CLASSIC STRIPE CARDIGAN

MEASUREMENTS

Three sizes	1st size	2nd size	3rd size
To fit bust	86-91cm/34-36in	91-97cm/36-38in	97-102cm/38-40in
Actual bust measurement	98cm/38½in	104cm/41in	110cm/43¼in
Length	73.5cm/29in	76cm/30in	78.5cm/31in
Sleeve seam	43cm/17in	43cm/17in	43cm/17in

MATERIALS

Yarn

Any **double-knit** weight cotton yarn can be used as long as it knits up to the given tension.
550(575:600)g/20(21:21)oz black (A)
350(375:400)g/13(14:15)oz white (B)

Needles and other materials

1 pair each 3¼mm (US 3) and 4mm (US 5) needles
4 buttons

TENSION

20 sts and 27 rows to 10cm/4in on 4mm (US 5) needles over st st.

BACK

With 3¼mm (US 3) needles and A, cast on 98(104:110) sts. Work 6 rows K1, P1 rib.
Change to 4mm (US 5) needles. Beg with a K row cont in st st, work 10 rows.
Change to B, work 16 rows.
Change to A, work 16 rows. These 32 rows form the stripe sequence.
Rep these 32 rows until work measures 73.5(76:78.5)cm/29(30:31)in from beg, ending with a P row.

Shape shoulders

Keeping stripe sequence correct, cast off 11(12:13) sts at beg of next 6 rows.
Cast off rem 32 sts.

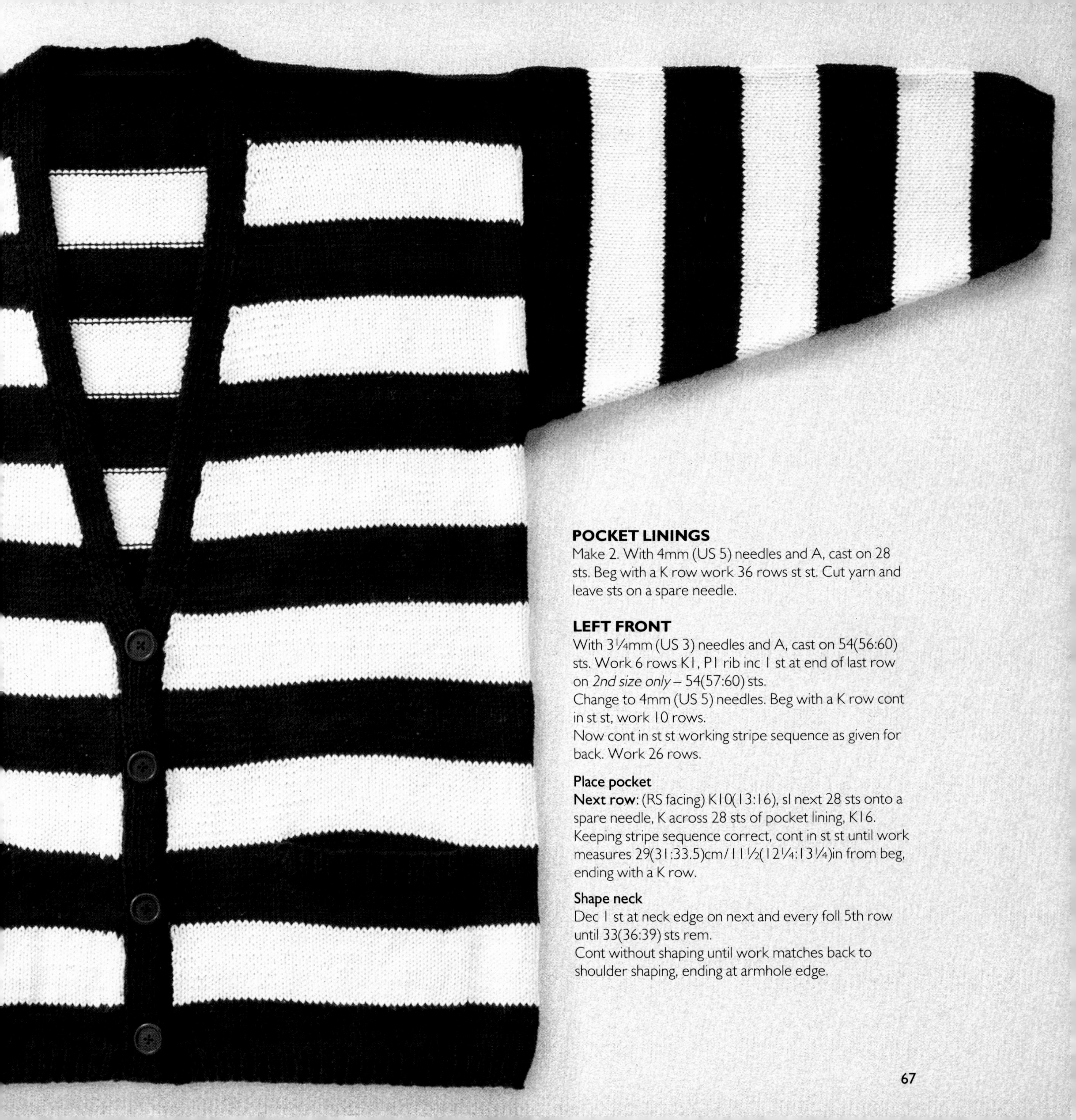

POCKET LININGS

Make 2. With 4mm (US 5) needles and A, cast on 28 sts. Beg with a K row work 36 rows st st. Cut yarn and leave sts on a spare needle.

LEFT FRONT

With 3¼mm (US 3) needles and A, cast on 54(56:60) sts. Work 6 rows K1, P1 rib inc 1 st at end of last row on *2nd size only* – 54(57:60) sts.
Change to 4mm (US 5) needles. Beg with a K row cont in st st, work 10 rows.
Now cont in st st working stripe sequence as given for back. Work 26 rows.

Place pocket

Next row: (RS facing) K10(13:16), sl next 28 sts onto a spare needle, K across 28 sts of pocket lining, K16.
Keeping stripe sequence correct, cont in st st until work measures 29(31:33.5)cm/11½(12¼:13¼)in from beg, ending with a K row.

Shape neck

Dec 1 st at neck edge on next and every foll 5th row until 33(36:39) sts rem.
Cont without shaping until work matches back to shoulder shaping, ending at armhole edge.

Shape shoulder
Cast off 11(12:13) sts at beg of next and foll alt row. Work 1 row. Cast off rem 11(12:13) sts.

RIGHT FRONT

Work as given for left front, reversing placing of pocket and all shapings.

SLEEVES

Make 2. With 3¼mm (US 3) needles and A, cast on 60 sts. Work 6 rows K1, P1 rib.
Change to 4mm (US 5) needles. Beg with a K row cont in st st, work 10 rows then cont in stripe sequence as given for back, *at the same time*, inc 1 st at each end of 3rd and every foll 4th row until there are 110 sts.
Cont without further shaping until work measures 43cm/17in from beg, ending with a P row.
Cast off loosely.

FRONT BAND

Join shoulder seams
With 3¼mm (US 3) needles and A, cast on 9 sts. Cont in K1, P1 rib as foll:
Row 1: (RS) K1, *P1, K1, rep from * to end.
Row 2: P1, *K1, P1, rep from * to end.
Rep these 2 rows until work measures 2.5cm/1in from beg, ending with a WS row.
1st buttonhole row: (RS) Rib 3, cast off 3, rib to end.
2nd buttonhole row: Rib, casting on 3 sts over those cast off in previous row.
Cont in rib as set, *at the same time*, making 3 more buttonholes 8(9:10)cm/3(3½:4)in above base of previous one, until band is long enough, when slightly stretched, to fit up right front edge, across back neck and down left front edge. Cast off in rib.

POCKET TOPS

Alike. With RS of work facing, 3¼mm (US 3) needles and A, return to sts of pocket on spare needle. Work 6 rows K1, P1 rib. Cast off in rib.

TO MAKE UP

Carefully press pieces, avoiding ribbing, using a steam iron or press under a damp cloth.
Placing centre of cast-off edge of sleeve to shoulder seam, set in sleeves.
Join side and sleeve seams. Catch down pocket linings on WS and pocket tops at side edges on RS. Sew on front band and buttons.

STRIPE AND BLOCK SWEATER

MEASUREMENTS

Two sizes	1st size	2nd size
To fit bust	81-86cm/32-34in	86-91cm/34-36in
Actual bust measurement	103.5cm/40¾in	109cm/43in
Length	54.5cm/21½in	57cm/22½in
Sleeve seam	35cm/13¾in	35cm/13¾in

MATERIALS

Yarn

Any **double-knit** weight cotton yarn can be used as long as it knits up to the given tension.
375(400)g/14(15)oz blue (A)
325(350)g/12(13)oz white (B)
50(50)g/2(2)oz yellow (C)
50(50)g/2(2)oz pink (D)

Needles

1 pair each 3¼mm (US 3) and 4mm (US 5) needles

TENSION

22 sts and 28 rows to 10cm/4in on 4mm (US 5) needles over st st.

Note
When working the stripe and block sections, wind off small amounts of each colour so that colour areas can be worked separately, twisting yarns around each other on wrong side at joins to avoid holes.

BACK

With 3¼mm (US 3) needles and A, cast on 114(120) sts and work 18 rows K1, P1 rib.
Change to 4mm (US 5) needles and B. Commence patt.
Beg with a K row, work 20 rows st st.
Next row: K11(14)A, 16C, 16D, 71(74)A.
Next row: P71(74)A, 16D, 16C, 11(14)A.
Rep last 2 rows 9 times more.
Next row: K11(14)B, 16D, 16C, 71(74)B.
Next row: P71(74)B, 16C, 16D, 11(14)B.
Rep last 2 rows 9 times more.
Change to A, work 20 rows st st.
Next row: K74B, 16C, 16D, 14B.
Next row: P14B, 16D, 16C, 71(74)B.
Rep last 2 rows 9 times more.
Next row: K71(74)A, 16D, 16C, 14A.
Next row: P14A, 16C, 16D, 71(74)A.
Rep last 2 rows 9 times more.
Cont in st st in stripes of 20 rows B, 20 rows A until work measures 54.5(57)cm/21½(22½)in from beg, ending with a P row.

Shape shoulders
Cast off 9(10) sts at beg of next 6 rows – 60 sts.

Commence collar
Cont in A, K 1 row.
Change to 3¼mm (US 3) needles and work 20 rows K1, P1 rib.
Cast off loosely in rib.

FRONT
Work as given for back.

SLEEVES
Make 2. With 3¼mm (US 3) needles and A, cast on 72 sts and work 16 rows K1, P1 rib.
Change to 4mm (US 5) needles.
Beg with a K row cont in st st in stripes of 20 rows B, 20 rows A, *at the same time*, inc 1 st at each end of 2nd and every foll 3rd row until there are 110 sts.
Cont without further shaping until work measures 35cm/13¾in from beg, ending with a P row.
Cast off loosely.

TO MAKE UP
Press lightly on wrong side of work, avoiding ribbing, using a steam iron or press under a damp cloth. Join shoulder seams and collar.
Placing centre of cast-off edge of sleeve to shoulder seams, set in sleeves. Join side and sleeve seams, matching stripes.

BIRDS AND BEASTS

Some of the more intriguing creatures of nature inspire the motifs on these sweaters.

On the **Birds of Paradise Sweater** exotic birds fly across a soft turquoise 'sky', and the pretty **Shell and Lace Sweater** has lacy bands alternating with subtle shell and starfish motifs knitted in four-ply yarn. Sunny yellow is used as the background for a charming butterfly motif which edges the **Butterfly Cardigan**. For the **Tropical Fish Cardigan**, vividly-coloured fish breathing silver bubbles from their mouths dart about a deep blue 'ocean'. Sea-horses, crabs, starfish and shells are scattered over the grey-green background of the **Sea Creatures Jacket** striped with white 'wavy' lines to add to the marine effect.

BIRDS OF PARADISE SWEATER

MEASUREMENTS

Three sizes	1st size	2nd size	3rd size
To fit bust	86-91cm/34-36in	91-97cm/36-38in	97-102cm/38-40in
Actual bust measurement	99cm/39in	105cm/41¼in	110cm/43¼in
Length	58.5cm/23in	60cm/23½in	61cm/24in
Sleeve seam	44.5cm/17½in	45.5cm/18in	47cm/18½in

MATERIALS

Yarn

Any **double-knit** weight yarn can be used as long as it knits up to the given tension.
525(550:575)g/19(20:21)oz turquoise (A)
50(50:75)g/2(2:3)oz blue (B)
50(50:50)g/2(2:2)oz green (C)
25(25:50)g/1(1:2)oz pink (D)
50(50:50)g/2(2:2)oz coral (E)
50(50:50)g/2(2:2)oz brown (F)
25(50:50)g/1(2:2)oz yellow (G)

Needles

1 pair each 3¼mm (US 3) and 4mm (US 5) needles

TENSION

21 sts and 28 rows to 10cm/4in on 4mm (US 5) needles over patt.

> **Note**
> • When reading charts, work K rows (odd-numbered rows) from right to left and P rows (even-numbered rows) from left to right.
> • When working the bird motifs, wind off small amounts of the required colours so that each colour area can be worked separately, twisting yarns around each other on wrong side at joins to avoid holes.

BACK

**With 3¼mm (US 3) needles and A, cast on 84(90:96) sts. Cont in K1, P1 rib for 10cm/4in.
Increase row: Rib and inc 20 sts evenly across row – 104(110:116) sts.
Change to 4mm (US 5) needles. Beg with a K row cont in st st working from back chart between appropriate lines for size required.**

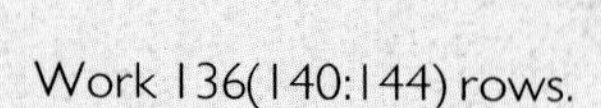

Work 136(140:144) rows.

Shape shoulders
Follow chart – cast off 11(12:13) sts at beg of next 6 rows. Cast off rem 38 sts.

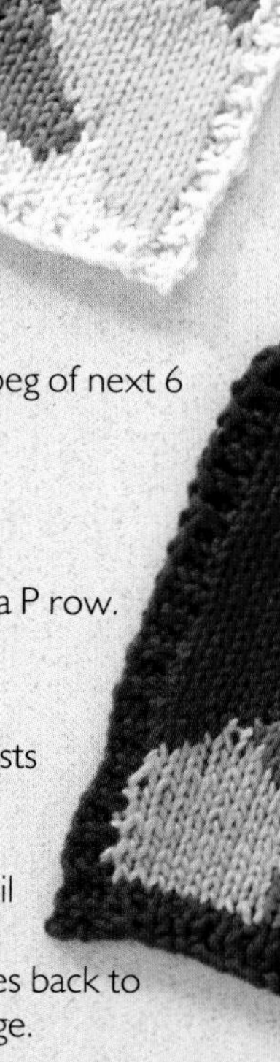

FRONT

Work as given for back from ** to **.
Work 116(120:124) rows, ending with a P row.

Shape neck
Next row: (RS facing – foll chart)
Patt 42(45:48) sts and turn, leaving rem sts on a spare needle.
Complete left side of neck first.
Dec 1 st at neck edge on every row until 33(36:39) sts rem.
Cont without shaping until work matches back to shoulder shaping, ending at armhole edge.

BACK AND FRONT CHART

Shape shoulder

Cast off 11(12:13) sts at beg of next and foll alt row. Work 1 row. Cast off rem 11(12:13) sts.
With RS of work facing, return to sts on spare needle, rejoin yarn, cast off centre 20 sts, patt to end. Complete as given for first side of neck.

SLEEVES

Make 2. With 3¼mm (US 3) needles and A, cast on 40 sts. Cont in K1, P1 rib for 10cm/4in.
Increase row: Rib and inc 20 sts evenly across row – 60 sts.
Change to 4mm (US 5) needles. Beg with a K row, cont in st st working from sleeve chart, *at the same time*, inc 1 st at each end of every 3rd row until there are 110 sts. Cont without shaping until row 92 of chart has been completed. Now cont in st st and A only until work measures 44.5(45.5:47)cm/17½(18:18½)in from beg, ending with a P row. Cast off loosely.

NECKBAND

Join right shoulder seam.
With RS of work facing, 3¼mm (US 3) needles and A, K up 24 sts down left side of neck, 20 sts from centre front, 24 sts up right side of neck and 38 sts across back neck – 106 sts.
Work 2.5cm/1in in K1, P1 rib. Cast off loosely in rib.

TO MAKE UP

Carefully press pieces avoiding ribbing using a steam iron or press under a damp cloth.
Join left shoulder and neckband seam. Placing centre of cast-off edge of sleeve to shoulder seam, set in sleeves. Join side and sleeve seams.

SLEEVE CHART

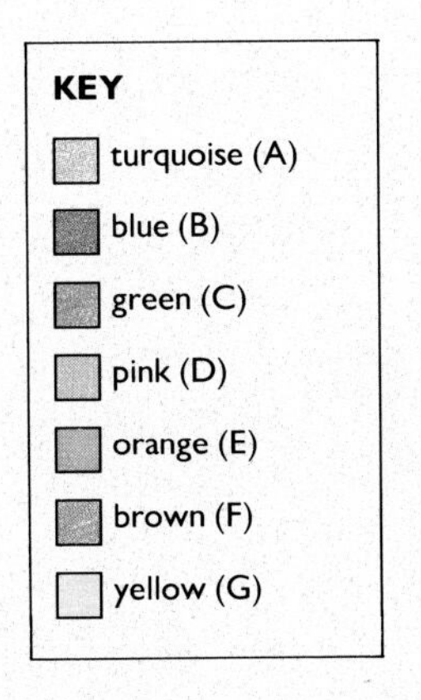

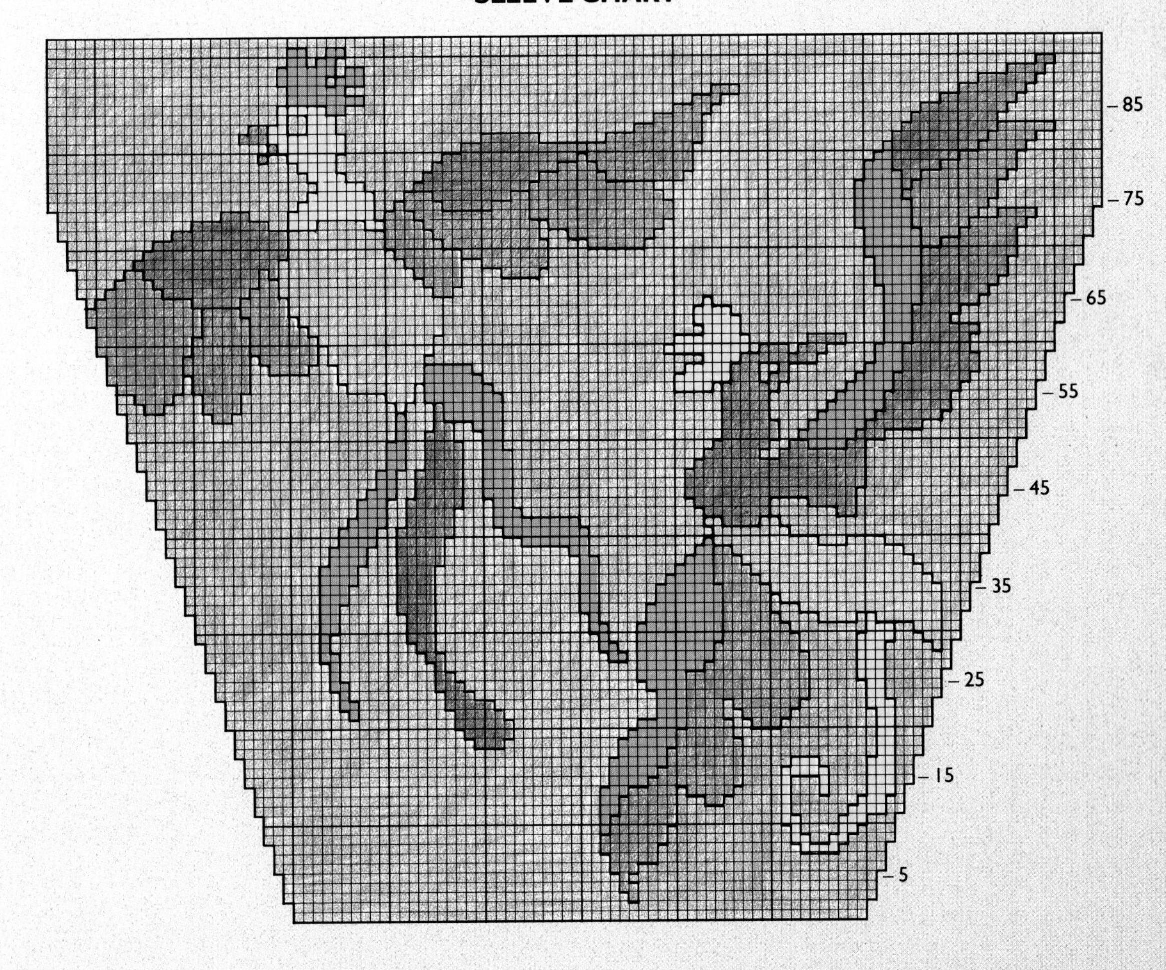

SHELL AND LACE SWEATER

MEASUREMENTS

One size	
To fit bust	86-102cm/34-40in
Actual bust measurement	110cm/43¼in
Length	55.5cm/21¾in
Sleeve seam	23cm/9in

MATERIALS

Yarn

Any **4-ply** weight cotton yarn can be used as long as it knits up to the given tension.
200g/8oz cream (A)
200g/8oz blue (B)
100g/4oz brown bouclé (C)
25g/1oz peach (D)
25g/1oz coral (E)
25g/1oz brown (F)

Needles

1 pair each 2¾mm (US 1) and 3¼mm (US 3) needles

TENSION

26 sts and 34 rows to 10cm/4in on 3¼mm (US 3) needles over patt.

Note

- When reading chart, work K rows (even-numbered rows) from right to left and P rows (odd-numbered rows) from left to right.
- When working the motifs, wind off small amounts of the required colours so that each colour area can be worked separately, twisting yarns around each other on wrong side at joins to avoid holes.

BACK

**With 2¾mm (US 1) needles and B, cast on 120 sts.
Cont in K1, P1 rib for 5cm (2in).
Increase row: Rib and inc 23 sts evenly across row – 143 sts.
Change to 3¼mm (US 3) needles and commence patt.
Change to A.
Row 1: (WS facing) P.
Row 2: *Sl 1, K1, psso, K3 tbl, yfwd, K1, yfwd, K3 tbl, K2 tog, rep from * to end.
Row 3: P.
Row 4: *Sl 1, K1, psso, K2 tbl, yfwd, K1, yfwd, sl 1, K1, psso, yfwd, K2 tbl, K2 tog, rep from * to end.
Row 5: P.
Row 6: *Sl 1, K1, psso, K1 tbl, yfwd, K1, yfwd, sl 1, K1, psso, yfwd, sl 1, K1, psso, yfwd, K1 tbl, K2 tog, rep from * to end.
Row 7: P.
Row 8: *Sl 1, K1, psso, yfwd, K1, yfwd, sl 1, K1, psso, yfwd, sl 1, K1, psso, yfwd, sl 1, K1, psso, yfwd, K2 tog, rep from * to end.
Row 9: P.
Change to C.
Rows 10-11: K.
Commence two-colour band in st st.
Row 12: *K3A, 1B, rep from * to last 3 sts, 3A.
Row 13: *P1B, 1A, rep from * to last st, 1B.
Row 14: K1A, *1B, 3A, rep from * to last 2 sts, 1B, 1A.
Row 15: With A, P.
Change to C.
Row 16: K and inc 5 sts evenly across row – 148 sts.
Row 17: K.
Rows 18-35: Beg with a K row cont in st st working from chart.
Change to C.
Row 36: K and dec 5 sts evenly across row – 143 sts.
Row 37: K.
Rows 38-41: Rep rows 12-15.
Change to C.
Rows 42-43: K.
Change to A.
Row 44: K.
These 44 rows form the patt. **Cont in patt until work measures approx 55.5cm/21¾in from beg, ending with row 39 of patt – 143 sts.

Shape shoulders

Cast off 16 sts at beg of next 6 rows. Cast off rem 47 sts.

FRONT

Work as given for back from ** to **.
Cont in patt until work measures approx 49cm/19¼in from beg, ending with row 17 of patt – 148 sts.

Shape neck

Next row: Patt 60 sts and turn leaving rem sts on a spare needle.

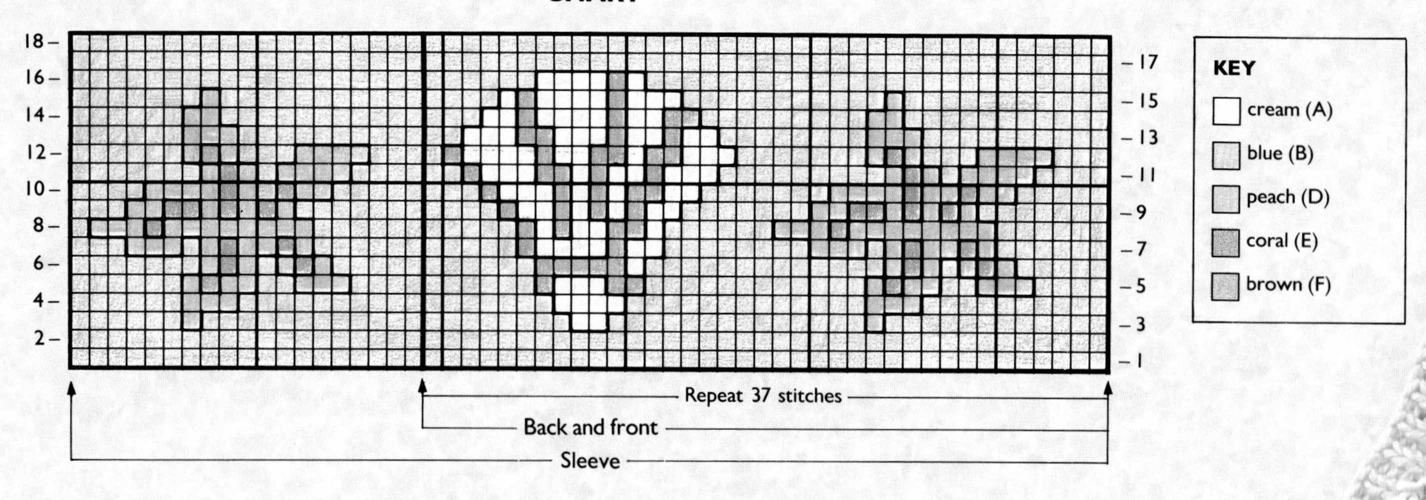

Complete left side of neck first.
Dec 1 st at neck edge on every row until 48 sts rem.
Cont without shaping (do not dec on row 36 of patt) until work matches back to shoulder shaping, ending at armhole edge.

Shape shoulder
Cast off 16 sts at beg of next and foll alt row. Work 1 row. Cast off rem 16 sts.
With RS of work facing, return to sts on spare needle, rejoin yarn, cast off centre 28 sts, patt to end.
Complete to match first side of neck.

SLEEVES

Make 2. With 2¾mm (US 1) needles and B, cast on 72 sts. Cont in K1, P1 rib for 5cm/2in.
Increase row: Rib and inc 60 sts evenly across row – 132 sts.
Change to 3¼mm (US 3) needles. Cont in patt as given for back but, *dec* 2 sts on row 16 and *inc* 2 sts on row 36.
Cont in patt until work measures approx 23cm/9in from beg, ending with row 15 of patt. Change to C.
K 2 rows.
Cast off loosely with C.

NECKBAND

Join right shoulder seam.
With RS of work facing, 2¾mm (US 1) needles and B, K up 32 sts down left side of neck, 28 sts from centre front, 32 sts up right side of neck and 48 sts from back neck – 140 sts.
Work in K1, P1 rib for 5cm/2in. Cast off loosely in rib.

TO MAKE UP

Carefully press pieces avoiding ribbing using a steam iron or press under a damp cloth.
Join left shoulder and neckband seam. Fold neckband in half onto RS and slip stitch in place. Placing centre of cast-off edge of sleeve to shoulder seams, set in sleeves. Join side and sleeve seams.

BUTTERFLY CARDIGAN

MEASUREMENTS

Two sizes	1st size	2nd size
To fit bust	86-91cm/34-36in	91-97cm/36-38in
Actual bust measurement	99cm/39in	105cm/41¼in
Length	49.5cm/19½in	50cm/19¾in
Sleeve seam	42.5cm/16¾in	42.5cm/16¾in

MATERIALS

Yarn

Any **double-knit** weight cotton yarn can be used as long as it knits up to the given tension.
625(650)g/22(23)oz yellow (A)
50(50)g/2(2)oz blue (B)
25(25)g/1(1)oz deep green (C)
25(25)g/1(1)oz turquoise (D)
25(25)g/1(1)oz brown (E)
25(25)g/1(1)oz coral (F)
50(50)g/2(2)oz cream (G)

Needles and other materials

1 pair each 3¼mm (US 3) and 4mm (US 5) needles
3 buttons

TENSION

21 sts and 28 rows to 10cm/4in on 4mm (US 5) needles over st st.

Note
- When reading charts, work K rows (odd-numbered rows) from right to left and P rows (even-numbered rows) from left to right.
- When working the motifs, wind off small amounts of the required colours so that each colour area can be worked separately, twisting yarns around each other on wrong side at joins to avoid holes.

LEFT FRONT

With 4mm (US 5) needles and A, cast on 2 sts. Beg with a K row, cont in st st working from left front chart. Foll chart, shape front point – inc 1 st at end (side edge) of 2nd and at same edge on every foll row, *at the same time*, inc 1 st at end (front edge) of 3rd row and at same edge on every alt row until there are 39 sts, ending with a P row.

Cont to inc 1st at front edge on next and 2 foll alt rows, *at the same time*, cast on 4 sts at beg (side edge) of next row and 3 sts at beg of foll 1(2) alt rows – 49(52) sts.
***Cont without shaping until row 78 of chart has been completed. Cont to foll chart, shape front neck – dec 1 st at front edge on next and every foll 4th row until 29(32) sts rem, then on every foll 3rd row until 26(28) sts rem, ending at armhole edge.

Shape shoulder

Still dec at front edge as before, cast off 8(9) sts at beg of next and foll alt row. Work 1 row. Cast off rem 9 sts.

FRONTS CHART

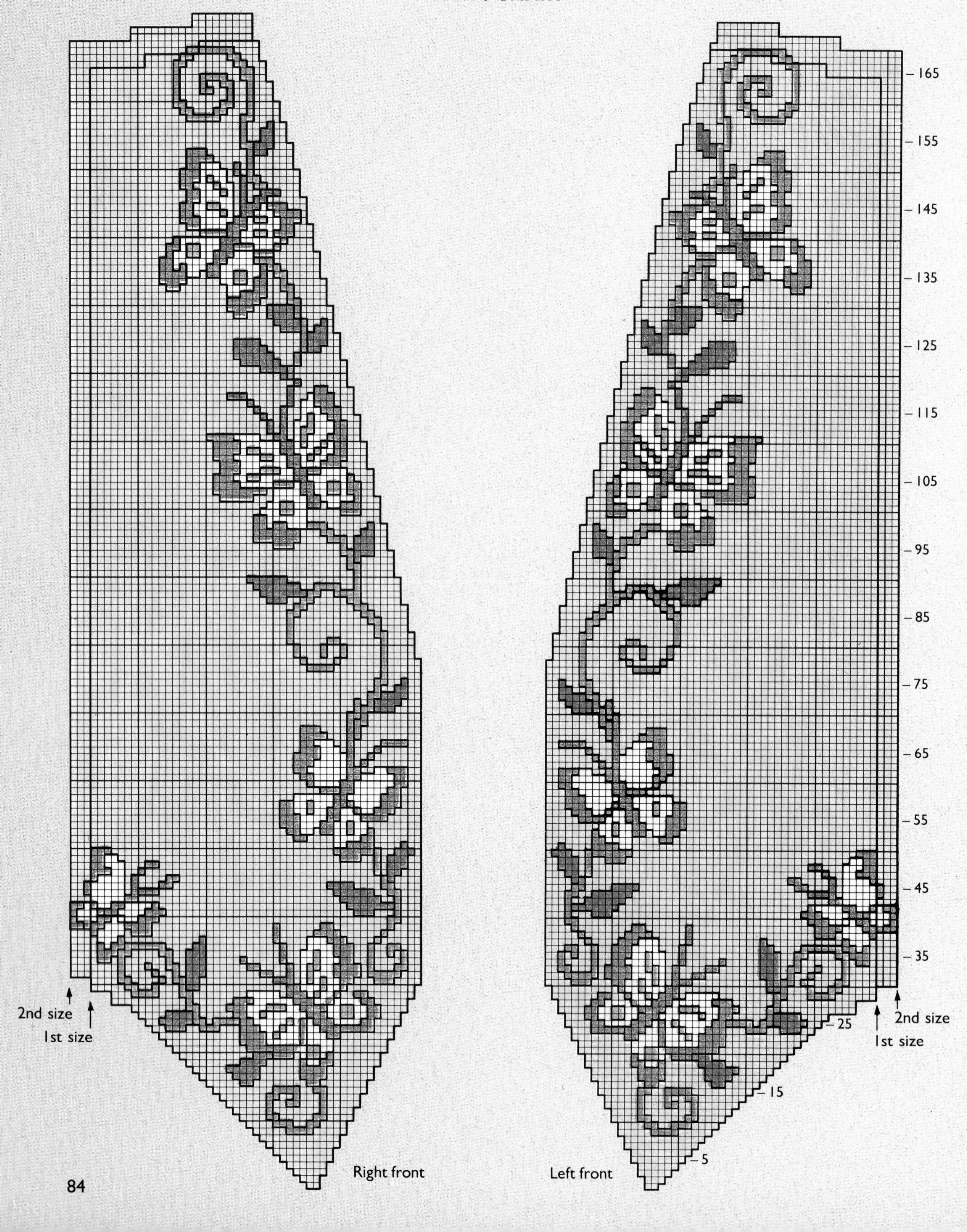

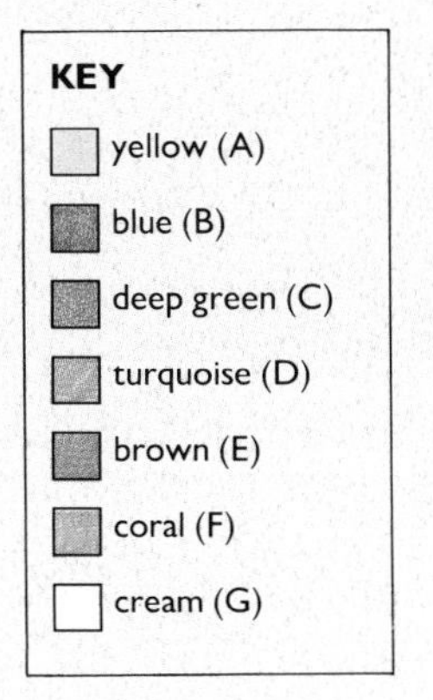

BACK AND SLEEVE CHART

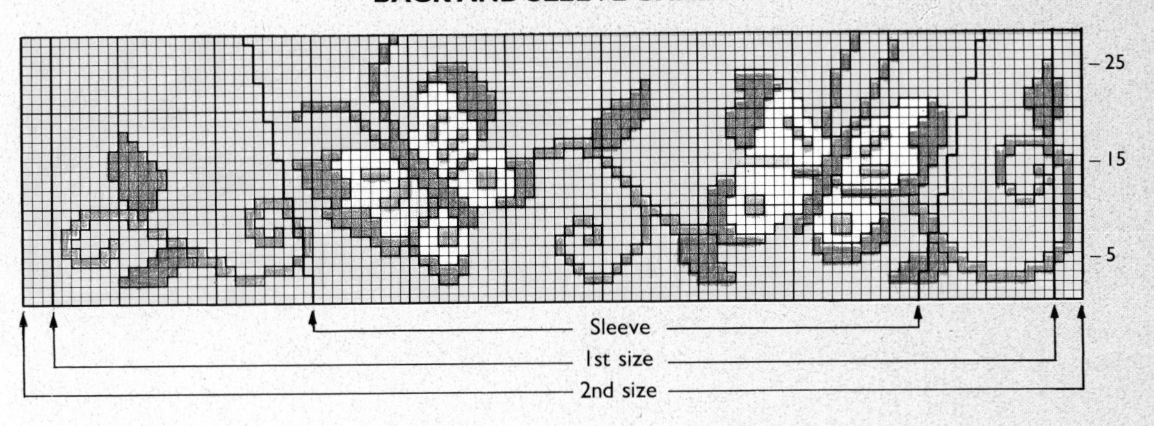

RIGHT FRONT

With 4mm (US 5) needles and A, cast on 2 sts. Beg with a K row, cont in st st working from right front chart. Foll chart, shape front point – inc 1 st at beg (side edge) of 2nd and at same edge on every foll row, *at the same time*, inc 1 st at beg (front edge) of 3rd row and at same edge on every alt row until there are 41 sts, ending with a K row.
Cont to inc 1 st at front edge on 2 foll alt rows, *at the same time*, cast on 3 sts at beg of next and foll 1(2) alt rows – 49(52) sts.
Complete to match left front from *** to end.

FRONT POINT EDGINGS

Front edges – alike

With RS of work facing, 3¼mm (US 3) needles and A, K up 28 sts from shaped front edge of point.
Work 2 rows K1, P1 rib. Cast off in rib.

Side edges – alike

With RS of work facing, 3¼mm (US 3) needles and A, K up 34(36) sts from shaped side edge of point.
Work 2 rows K1, P1 rib. Cast off in rib.

BACK

With 3¼mm (US 3) needles and A, cast on 104(110) sts. Cont in K1, P1 rib for 2.5cm/1in.
Change to 4mm (US 5) needles. Beg with a K row, cont in st st working from back chart between appropriate lines for size required. Work 28 rows.
Now cont in st st in A only until work measures same as left front from edging to shoulder shaping, ending with a P row.

Shape shoulders

Cast off 8(9) sts at beg of next 4 rows and 9 sts at beg of foll 2 rows.
Cast off rem 54(56) sts.

SLEEVES

Make 2. With 3¼mm (US 3) needles and A, cast on 60 sts. Cont in K1, P1 rib for 2.5cm/1in.
Increase row: Rib and inc 3 sts evenly across row – 63 sts.
Change to 4mm (US 5) needles. Beg with a K row cont in st st working from sleeve chart. Work 28 rows, *at the same time*, shape sleeve, foll chart – inc 1 st at each end of every 4th row – 77 sts.
Now cont in st st in A only, inc as before at each end of every foll 4th row until there are 101 sts.
Now cont without shaping until work measures 42.5cm/16¾in from beg, ending with a P row. Cast off loosely.

FRONT BAND

Join shoulder seams.
With 3¼mm (US 3) needles and A, cast on 9 sts.
Row 1: (RS facing) K1, *P1, K1, rep from * to end.
Row 2: P1, *K1, P1, rep from * to end.
Rep these 2 rows for 2.5cm/1in, ending with a WS row.
1st buttonhole row: (RS) Rib 3, cast off 3, rib to end.
2nd buttonhole row: Rib to end, casting on 3 sts over those cast off in previous row.
Cont in rib as set, *at the same time*, making 2 more buttonholes 7cm/2¾in above base of previous one, until band is long enough, when slightly stretched, to fit up right front edge, across back neck and down left front edge. Cast off in rib.

TO MAKE UP

Carefully press pieces, avoiding ribbing, using a steam iron or press under a damp cloth.
Placing centre of cast-off edge of sleeve to shoulder seam, set in sleeves.
Join side and sleeve seams. Catch together row ends of front point edgings at centre front. Sew on front band.
Sew on buttons.

TROPICAL FISH CARDIGAN

MEASUREMENTS

Two sizes	1st size	2nd size
To fit bust	81-86cm/32-34in	91-97cm/36-38in
Actual bust measurement	98.5cm/38¾in	104cm/41in
Length	59cm/23¼in	61cm/24in
Sleeve seam	45cm/17¾in	47cm/18½in

MATERIALS

Yarn

Any **double-knit** weight cotton yarn can be used as long as it knits up to the given tension.
650(700)g/23(25)oz deep blue (A)
25(25)g/1(1)oz each orange (B), lime green (C), silver (D), black (E), turquoise (F), green (G), pink (H), red (I), purple (J)

Needles and other materials

1 pair each 3¼mm (US 3) and 4mm (US 5) needles
8 glass buttons
80 silver beads

TENSION

21 sts and 27 rows to 10cm/4in on 4mm (US 5) needles over patt.

Note

- Use silver yarn (D) double throughout.
- When reading charts, work K rows (odd-numbered rows) from right to left and P rows (even-numbered rows) from left to right.
- When working the motifs, wind off small amounts of the required colours so that each motif can be worked separately, twisting yarns around each other on wrong side at joins to avoid holes.

FRONTS CHART

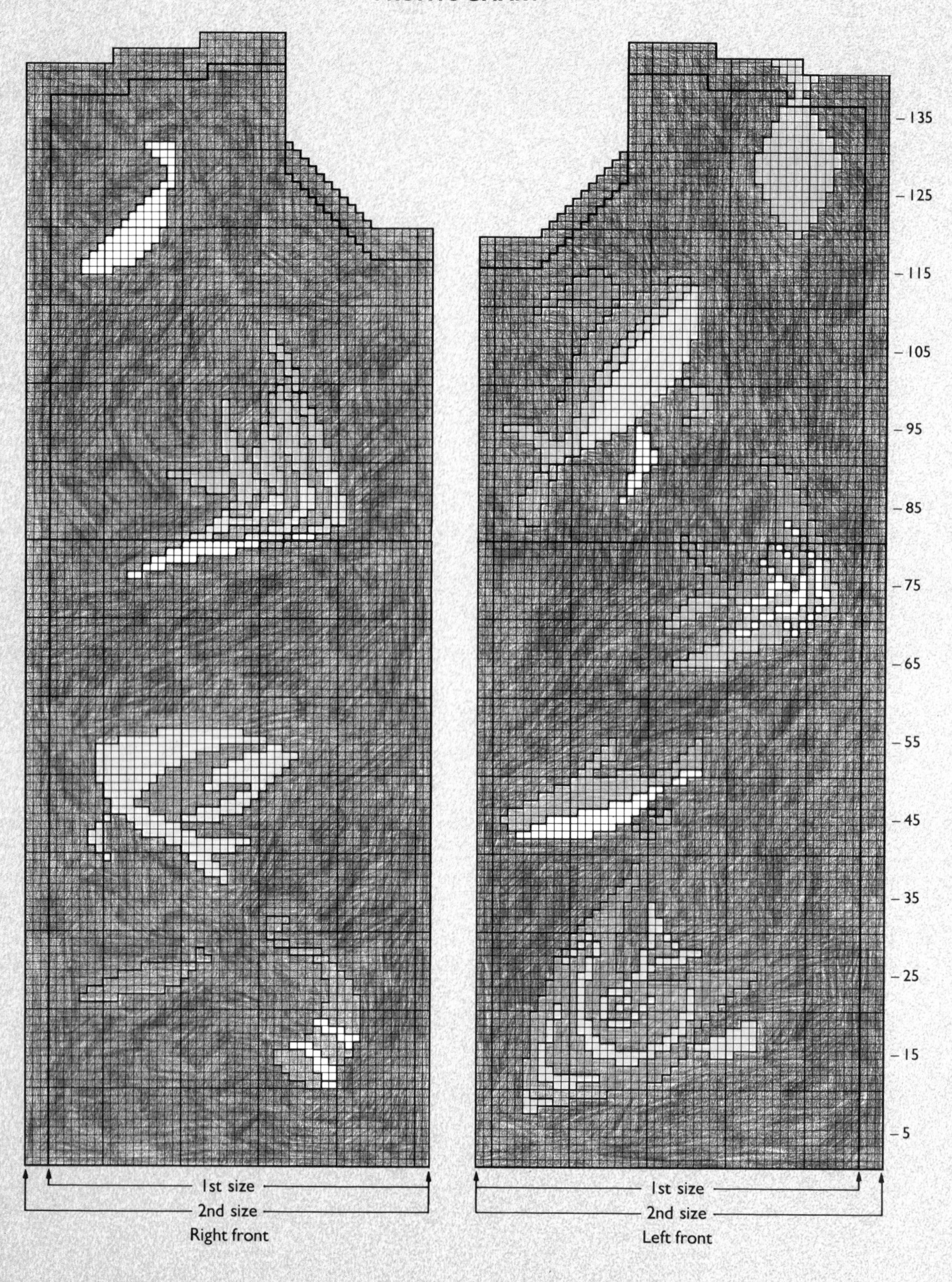

BACK CHART

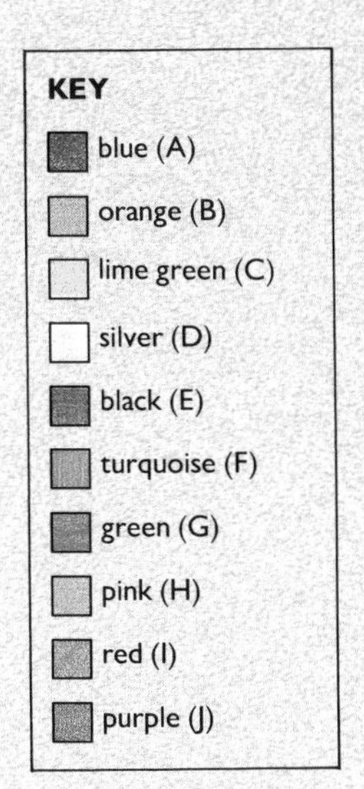
KEY
blue (A)
orange (B)
lime green (C)
silver (D)
black (E)
turquoise (F)
green (G)
pink (H)
red (I)
purple (J)

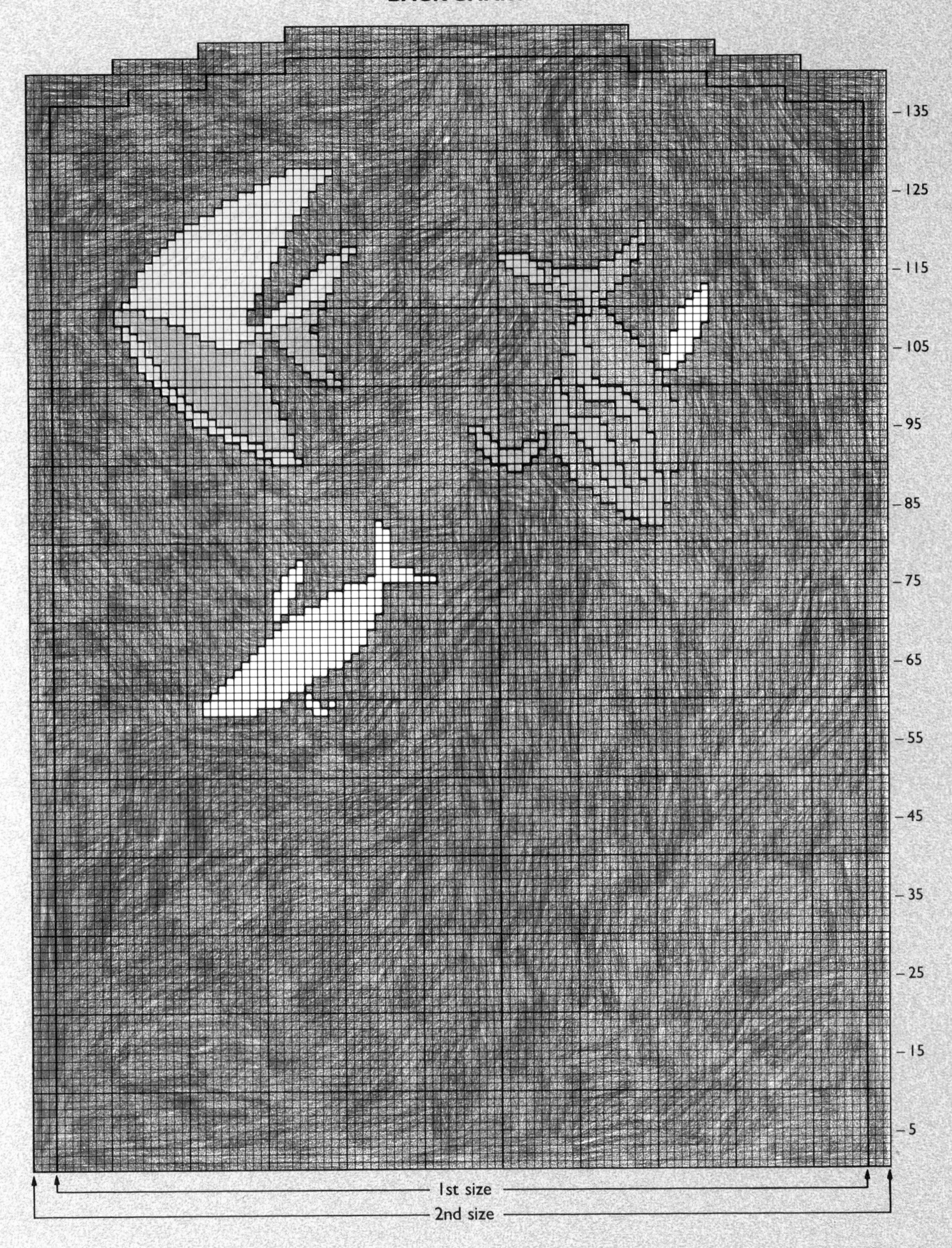
135
125
115
105
95
85
75
65
55
45
35
25
15
5
1st size
2nd size

SLEEVE CHART

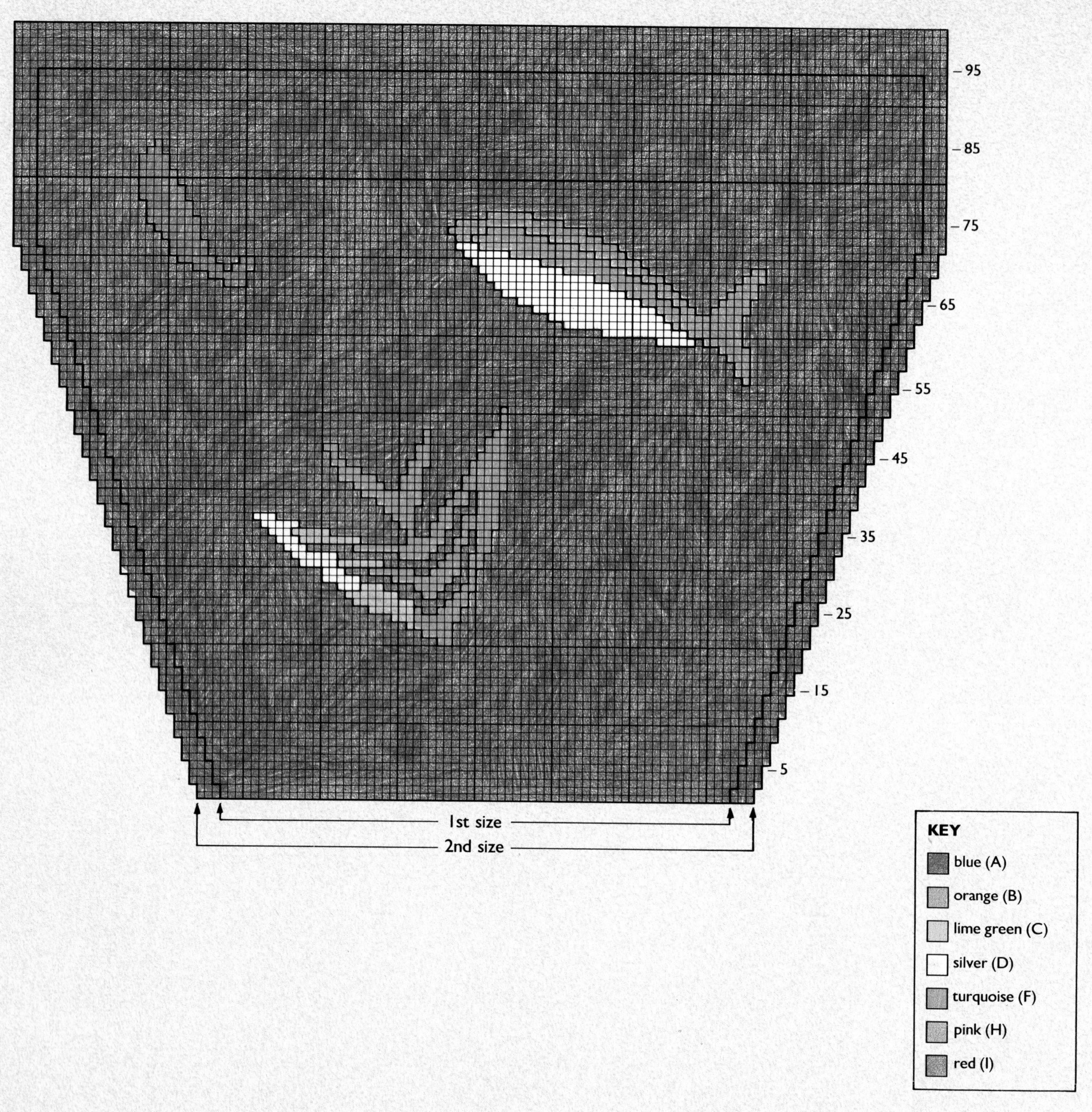

LEFT FRONT

With 3¼mm (US 3) needles and A, cast on 38(42) sts.
Cont in K2, P2 rib as foll:
Row 1: (RS facing) K2, *P2, K2, rep from * to end.
Row 2: P2, *K2, P2, rep from * to end.
Rep these 2 rows for 9cm/3½in, ending with a RS row.
Increase row: Rib and inc 11(10) sts evenly across row – 49(52) sts.
Change to 4mm (US 5) needles. Beg with a K row cont in st st working from left front chart between appropriate lines for size required.
Work 115(119) rows.

Shape neck

Next row: (WS facing – foll chart) Cast off 8 sts, patt to end.
Dec 1 st at neck edge on every row until 30(33) sts rem.
Cont without shaping until row 136(140) of chart has been completed.

Shape shoulder

Cast off 10(11) sts at beg of next and foll alt row. Work 1 row. Cast off rem 10(11) sts.

RIGHT FRONT

Work as given for left front, working from appropriate chart and reversing all shapings by working 1 more row before neck and shoulder shaping.

BACK

With 3¼mm (US 3) needles and A, cast on 84(88) sts.
Cont in K2, P2 rib for 9cm/3½in.
Increase row: Rib and inc 20(22) sts evenly across row – 104(110) sts.
Change to 4mm (US 5) needles. Beg with a K row cont in st st working from back chart between appropriate lines for size required.
Work 136(140) rows of chart.

Shape shoulders

Cast off 10(11) sts at beg of next 6 rows. Cast off rem 44 sts.

SLEEVES

Make 2. With 3¼mm (US 3) needles and A, cast on 42(46) sts. Cont in K2, P2 rib as given for back for 10cm/4in.
Increase row: Rib and inc 24(26) sts evenly across row – 66(72) sts.
Change to 4mm (US 5) needles. Beg with a K row cont in st st working from sleeve chart between appropriate lines for size required. Foll chart, inc 1 st at each end of every 3rd row until there are 114(120) sts.
Cont without shaping until row 94(100) of chart has been completed.
Cast off loosely.

NECKBAND

Join shoulder seams.
With RS of work facing, 3¼mm (US 3) needles and A, K up 27 sts up right front neck, 42 sts from back neck and 27 sts down left front neck – 96 sts.
Work in K2, P2 rib for 5cm/2in. Cast off loosely in rib.

BUTTONHOLE BAND

Fold neckband in half onto WS and slip stitch in place.
With RS of work facing, 3¼mm (US 3) needles and A, K up 134(138) sts evenly up right front edge and neckband.
Beg with row 2, work 2 rows K2, P2 rib as given for back.
1st buttonhole row: Rib 3(4), cast off 3 sts, rib 6(7) including st used to cast off, *cast off 3 sts, rib 18 including st used to cast off, rep from * 4 times more, cast off 3 sts, rib 8(9) including st used to cast off, cast off 3 sts, rib to end.
2nd buttonhole row: Rib to end, casting on 3 sts over those cast off in previous row.
Work 2 rows rib.
Cast off in rib.

BUTTON BAND

With RS of work facing, 3¼mm (US 3) needles and A, K up 134(138) sts evenly up left front edge and neckband.
Beg with row 2, work 6 rows rib as given for back. Cast off in rib.

TO MAKE UP

Press lightly on wrong side of work avoiding ribbing using a steam iron or press under a damp cloth.
Placing centre of cast-off edge of sleeves to shoulder seams, set in sleeves. Join side and sleeve seams.
Sew silver beads as 'bubbles' from each fish mouth as shown in the photograph.
Sew on buttons.

SEA CREATURES JACKET

MEASUREMENTS

Three sizes	1st size	2nd size	3rd size
To fit bust	86-91cm/34-36in	91-97cm/36-38in	97-102cm/38-40in
Actual bust measurement	106cm/41¾in	113cm/44½in	120cm/47¼in
Length	74cm/29¼in	75cm/29½in	77cm/30½in
Sleeve seam	43cm/17in	47cm/18½in	50cm/19¾in

MATERIALS

Yarn

Any **double-knit** weight cotton yarn can be used as long as it knits up to the given tension.
725(750:775)g/26(27:28)oz grey/green (A)
125(125:150)g/5(5:6)oz cream (B)
50(50:50)g/2(2:2)oz pale blue (C)
25(25:25)g/1(1:1)oz turquoise (D)
25(25:25)g/1(1:1)oz dark turquoise (E)
25(25:25)g/1(1:1)oz dark pink (F)
50(50:50)g/2(2:2)oz pale pink (G)
25(25:25)g/1(1:1)oz coral (H)
50(50:50)g/2(2:2)oz yellow (I)

Needles and other materials

1 pair each 3¼mm (US 3) and 4mm (US 5) needles
3¼mm (US 3) circular needle
7 buttons

TENSION

22½ sts and 25 rows to 10cm/4in on 4mm (US 5) needles over patt.

> **Note**
> • When reading charts, work K rows (odd-numbered rows) from right to left and P rows (even-numbered rows) from left to right.
> • When working the motifs, wind off small amounts of the required colours so that each colour area can be worked separately, twisting yarns around each other on wrong side at joins to avoid holes.
> • When working the wavy lines yarn can be carried on wrong side of work over not more than 3 stitches at a time to keep fabric elastic.

BACK

With 3¼mm (US 3) needles and A, cast on 120(128:136) sts. Cont in K1, P1 rib for 5cm/2in.
Change to 4mm (US 5) needles. Beg with a K row cont in st st working from back chart between appropriate lines for size required.
Work 26 rows.

Commence pocket linings

Row 27: (Foll chart RS facing) With A, cast on 24 sts, K across these sts, work from chart to end.
Row 28: With A, cast on 24 sts, P across these sts, work from chart to last 24 sts, P to end in A. Keeping 24 sts in A and st st for pocket linings at each end of row and chart patt correct, work until row 54 of chart has been completed.

Shape top of linings

Now cast off 4 sts at beg of next 12 rows – 120(128:136) sts.
Cont to work from chart until row 172(176:180) has been completed, ending with a P row.

Shape shoulders

Cast off 13(13:15) sts at beg of next 2 rows and 12(14:15) sts at beg of foll 4 rows.
Cast off rem 46 sts.

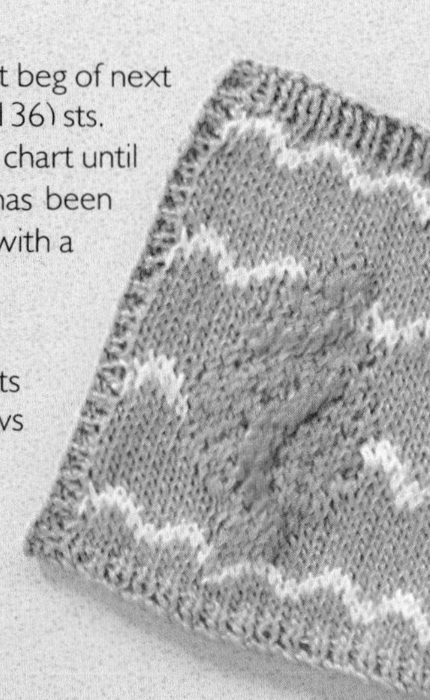

BACK, FRONTS AND SLEEVE CHART

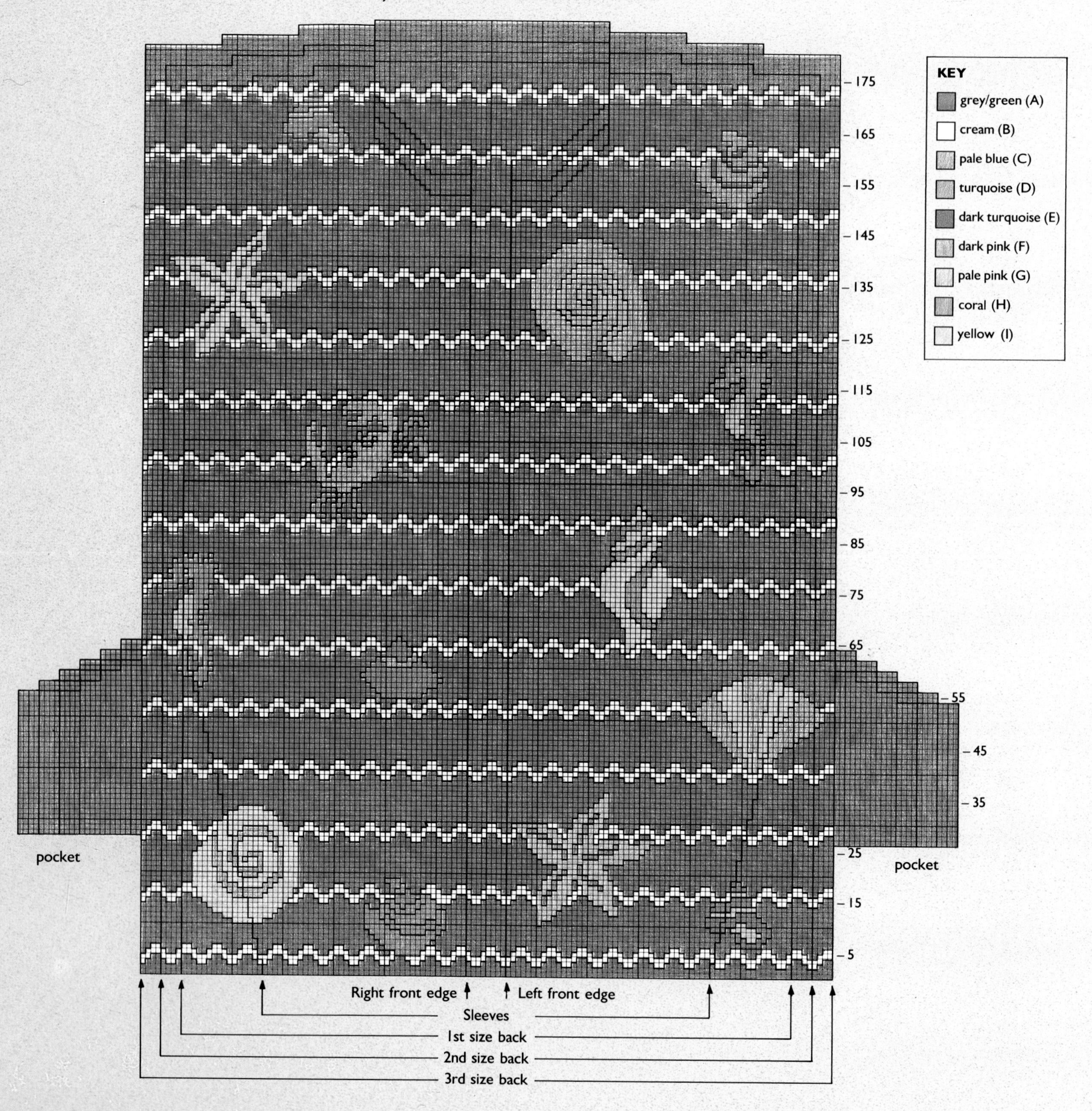

LEFT FRONT

**With 3¼mm (US 3) needles and A, cast on 56(60:64) sts. Cont in K1, P1 rib for 5cm/2in.
Change to 4mm (US 5) needles. Beg with a K row, cont in st st working from chart between front edge marker and appropriate lines for size required at back side edge.**
Work 26 rows.

Commence pocket lining

Row 27: (Foll chart – RS facing) With A, cast on 24 sts, K across these sts, work from chart to end.
Row 28: Work from chart to last 24 sts, P to end in A.
***Keeping 24 sts at side edge in st st and A and chart patt correct, work until row 54 of chart has been completed.

Shape top of lining

Cast off 4 sts at beg of next and 5 foll alt rows – 56(60:64) sts.
Cont working from chart until row 151(155:159) has been worked, ending with a K row.

Shape neck

Follow chart – cast off 8 sts at beg of next row, then dec 1 st at neck edge on every row until 37(41:45) sts rem.
Cont without shaping until row 172(176:180) of chart has been worked, ending with a P row.

Shape shoulder

Cast off 13(13:15) sts at beg of next row and 12(14:15) sts at beg of foll alt row. Work 1 row. Cast off rem 12(14:15) sts.

RIGHT FRONT

Work as given for left front from ** to **.
Work 27 rows.

Commence pocket lining

Row 28: (Foll chart – WS facing) With A, cast on 24 sts, P across these sts, work from chart to end.
Row 29: Work from chart to last 24 sts, K to end in A.
Complete as given for left front from *** to end, reversing all shapings so working 1 more row before top of lining, neck and shoulder shapings.

SLEEVES

Make 2. With 3¼mm (US 3) needles and A, cast on 44 sts. Cont in K1, P1 rib for 5cm/2in.
Increase row: Rib and inc into every st across row – 88 sts.
Change to 4mm (US 5) needles. Beg with a K row cont in st st working from chart between appropriate markers for sleeve, *at the same time*, inc 1 st at each end of every 4th row until there are 120 sts.
Cont without further shaping until row 96(104:112) of chart has been completed. Cast off loosely.

BUTTONHOLE BAND

With RS of work facing, 3¼mm (US 3) circular needle and A, K up 164(168:172) sts evenly up right front edge.
Work 3 rows K1, P1 rib.
1st buttonhole row: Rib 2(4:6), *cast off 4 sts, rib 22 including st used to cast off, rep from * 5 times more, cast off 4 sts, rib to end.
2nd buttonhole row: Rib to end, casting on 4 sts over those cast off in previous row.
Work 3 rows rib. Cast off in rib.

BUTTON BAND

With RS of work facing, 3¼mm (US 3) circular needle and A, K up 164(168:172) sts evenly down left front edge.
Work 8 rows K1, P1 rib. Cast off in rib.

NECKBAND

Join shoulder seams.
With RS of work facing, 3¼mm (US 3) needles and A, beg at centre of buttonhole band, K up 32 sts up right side of neck, 40 sts from across back neck and 32 sts down left side of neck to centre of button band – 104 sts.
Work 12 rows K1, P1 rib.
Cast off loosely in rib.

TO MAKE UP

Carefully press pieces avoiding ribbing, using a steam iron or press under a damp cloth.
Placing centre of cast-off edges of sleeves to shoulder seams, set in sleeves.
Join side seams, including edges of pockets and sleeve seams.
Catch the pocket lining on WS of fronts to hold in place.
Sew on buttons.

LIST OF SUPPLIERS

UK YARN SUPPLIERS

Colourtwist Ltd
10 Mayfield Avenue Ind. Park
Weyhill
Andover
Hants SP11 8HU
tel: (026 477) 3369

Brinion Heaton Ltd
The Old School
Queen Square
Saltford
Bristol BS18 3EL
tel: (02217) 3142

Naturally Beautiful Ltd
Main Street
Dent
Cumbria LA10 5QL
tel: (05875) 421

Rowan Yarns
Green Lane Mill
Holmfirth
West Yorkshire HD7 1RW
tel: (0484) 686714/687374

Smallwares Ltd
17 Galena Road
King Street
Hammersmith
London W6 0LU
tel: (01) 748 8511

Texere Yarns
College Mill
Barkerend Road
Bradford BD3 9AQ
tel: (0274) 22191

Pamela Wise
Old Loom House
Back Church House
London E1 1LS
tel: (01) 488 0144

UK RETAIL OUTLETS

Colourway
112a Westbourne Grove
London W2 5RU
tel: (01) 229 1432

Creativity
45 New Oxford Street
London WC1
tel: (01) 240 2945

Ries Wools
242 High Holborn
London WC1V 7DZ
tel: (01) 242 7721

Shepherds Purse
2 John Street
Bath BA1 2JL
tel: (0225) 310790

BUTTONS

The Button Box
44 Bedford Street
Covent Garden
London WC2E 9HA
tel: (01) 240 2716/2841
(send SAE for catalogue)

USA AND CANADA YARN SUPPLIERS

Noro Yarns
Agent: Mr Sion Elaluf
Knitting Fever
180 Babylon Turnpike
Roosevelt
New York, NY 11575
tel: (516) 546 3600
(800) 645 3457

Rowan Yarns (USA & Canada)
Agents:
Westminster Trading
5 Northern Boulevard
Amherst
New Hampshire 03031
tel: (603) 886 5041

William Unger & Co Inc
P.O. Box 1621
2478 Main Street
Bridgeport
Connecticut 06601
tel: (203) 335 5000

Estelle Designs and Sales Ltd
38 Continental Place
Scarborough
Ontario
Canada M1R 2TA
tel: (416) 2989922

USA RETAIL OUTLETS

Fiber Works
313 East 45th Street
New York, NY 10017
tel: (212) 286 9116

School Products Co Inc
1201 Broadway
New York, NY 10001
tel: (212) 679 3516

BUTTONS

Tender Buttons
143 East 62nd Street
New York, NY 10021